Environmental Epidemiology

ADVANCES IN CHEMISTRY SERIES 241

Environmental Epidemiology

Effects of Environmental Chemicals on Human Health

William M. Draper, EDITOR
California Department of Health Services

Developed from a symposium sponsored
by the Division of Environmental Chemistry, Inc.
at the 203rd National Meeting
of the American Chemical Society,
San Francisco, California,
April 5–10, 1992

American Chemical Society, Washington, DC 1994

Library of Congress Cataloging-in-Publication Data

Environmental epidemiology: effects of environmental chemicals on human health / William M. Draper, editor.
  p.   cm.—(Advances in chemistry series, ISSN 0065-2393; 241).

"Developed from a symposium sponsored by the Division of Environmental Chemistry, Inc., at the 203rd National Meeting of the American Chemical Society, San Francisco, California, April 5–10, 1992."

Includes bibliographical references and index.

ISBN 0–8412–2517–6 (clothbound) ISBN 0–8412–2933–3 (paperback)
  1. Environmental health—Congresses. 2. Pollution—Toxicology—Congresses.
  3. Environmentally induced diseases—Epidemiology—Congresses.

  I. Draper, William M., 1952– . II. American Chemical Society. Division of Environmental Chemistry, Inc. III. American Chemical Society. Meeting (203rd: 1992: San Francisco, Calif.). IV. Series

QDi.A355 no. 241
[RA565]
540 s—dc20   94–28456
[615.9(fm02]   CIP

The paper used in this publication meets the minimum requirements of American National Standard for Information Sciences—Permanence of Paper for Printed Library Materials, ANSI Z39.48–1984.

Copyright © 1994

American Chemical Society

All Rights Reserved. The appearance of the code at the bottom of the first page of each chapter in this volume indicates the copyright owner's consent that reprographic copies of the chapter may be made for personal or internal use or for the personal or internal use of specific clients. This consent is given on the condition, however, that the copier pay the stated per-copy fee through the Copyright Clearance Center, Inc., 27 Congress Street, Salem, MA 01970, for copying beyond that permitted by Sections 107 or 108 of the U.S. Copyright Law. This consent does not extend to copying or transmission by any means—graphic or electronic—for any other purpose, such as for general distribution, for advertising or promotional purposes, for creating a new collective work, for resale, or for information storage and retrieval systems. The copying fee for each chapter is indicated in the code at the bottom of the first page of the chapter.

The citation of trade names and/or names of manufacturers in this publication is not to be construed as an endorsement or as approval by ACS of the commercial products or services referenced herein; nor should the mere reference herein to any drawing, specification, chemical process, or other data be regarded as a license or as a conveyance of any right or permission to the holder, reader, or any other person or corporation, to manufacture, reproduce, use, or sell any patented invention or copyrighted work that may in any way be related thereto. Registered names, trademarks, etc., used in this publication, even without specific indication thereof, are not to be considered unprotected by law.

PRINTED IN THE UNITED STATES OF AMERICA

1994 Advisory Board

Advances in Chemistry Series

M. Joan Comstock, *Series Editor*

Robert J. Alaimo
Procter & Gamble Pharmaceuticals

Mark Arnold
University of Iowa

David Baker
University of Tennessee

Arindam Bose
Pfizer Central Research

Robert F. Brady, Jr.
Naval Research Laboratory

Margaret A. Cavanaugh
National Science Foundation

Arthur B. Ellis
University of Wisconsin at Madison

Dennis W. Hess
Lehigh University

Hiroshi Ito
IBM Almaden Research Center

Madeleine M. Joullie
University of Pennsylvania

Lawrence P. Klemann
Nabisco Foods Group

Gretchen S. Kohl
Dow-Corning Corporation

Bonnie Lawlor
Institute for Scientific Information

Douglas R. Lloyd
The University of Texas at Austin

Cynthia A. Maryanoff
R. W. Johnson Pharmaceutical Research Institute

Julius J. Menn
Western Cotton Research Laboratory, U.S. Department of Agriculture

Roger A. Minear
University of Illinois at Urbana–Champaign

Vincent Pecoraro
University of Michigan

Marshall Phillips
Delmont Laboratories

George W. Roberts
North Carolina State University

A. Truman Schwartz
Macalaster College

John R. Shapley
University of Illinois at Urbana–Champaign

L. Somasundaram
DuPont

Michael D. Taylor
Parke-Davis Pharmaceutical Research

Peter Willett
University of Sheffield (England)

FOREWORD

The ADVANCES IN CHEMISTRY SERIES was founded in 1949 by the American Chemical Society as an outlet for symposia and collections of data in special areas of topical interest that could not be accommodated in the Society's journals. It provides a medium for symposia that would otherwise be fragmented because their papers would be distributed among several journals or not published at all.

Papers are reviewed critically according to ACS editorial standards and receive the careful attention and processing characteristic of ACS publications. Volumes in the ADVANCES IN CHEMISTRY SERIES maintain the integrity of the symposia on which they are based; however, verbatim reproductions of previously published papers are not accepted. Papers may include reports of research as well as reviews, because symposia may embrace both types of presentation.

ABOUT THE EDITOR

WILLIAM M. DRAPER is a research chemist with the State of California Department of Health Services (CDHS). He received degrees in biochemistry (B.S., 1974) and agricultural and environmental chemistry (Ph.D., 1979) at the University of California at Davis. Later he spent two years as a postdoctoral researcher at the Pesticide Chemistry and Toxicology Laboratory at the Berkeley campus.

During the past 15 years his research interests in environmental chemistry included chemical fate and transport in aquatic systems, emissions testing in internal combustion engines used in underground mining, and human biomonitoring. He is particularly interested in trace organic chemical analysis by mass spectrometry and other instrumental techniques and in the measurement of reaction rate constants that determine chemical fate in the environment.

Dr. Draper has served on the faculties of Utah State University and the University of Minnesota School of Public Health. While at Utah State University he was chief chemist at the Epidemiologic Studies Program laboratory, with support from the U.S. Environmental Protection Agency Office of Pesticide Programs. In his current position his co-workers include epidemiologists, toxicologists, and physicians from California's Department of Health Services and Environmental Protection Agency.

CONTENTS

Preface xi

1. Assessing Environmental Risk—Scientifically Defensible or Fantasy? 1
   Vernon N. Houk

2. The Epidemiological Method 9
   Gary H. Spivey

3. Design Methods for Occupational and Environmental Epidemiology 21
   MaryFran Sowers

4. Interpreting Epidemiological Studies 29
   George Maldonado

5. Environmental Epidemiology for Chemists 39
   Curtis D. Klaassen and William C. Kershaw

6. Toxicology and Risk Assessment 53
   Arthur L. Craigmill, Scott Wetzlich, and William M. Draper

7. In Vitro and In Vivo Assays to Screen for Reproductive Toxicants in Animals and Humans 65
   Barbara S. Shane

8. The Gene-Tox Program: Data Evaluation of Chemically Induced Mutagenicity 89
   Michael C. Cimino and Angela E. Auletta

9. Biological, Biochemical, and Molecular Markers in Environmental Epidemiology 105
   Marilyn F. Vine

10. Examples of Measuring Internal Dose for Assessing Exposure in Epidemiological Studies 121
    Larry L. Needham

11. Research Strategy for Assessing Human Health Risks from Exposure to DNA-Reactive Chemicals: 1,3-Butadiene as a Case Study 137
    James A. Bond, Leslie Recio, and Roger O. McClellan

12. Molecular Epidemiology of Acrylonitrile: Indicators of Health Risk by Worker Surveillance and Regiospecific Modification of Ha-*ras* Oncogene 153
    John L. Wong, Bo Yuan, Peide Zhang, and Carlo H. Tamburro

13. Estimation of Risk of Kidney Dysfunction from Exposure to Cadmium Using Studies of Occupationally Exposed Workers 175
    Elizabeth A. Grossman and Caroline S. Freeman

14. Estimating Malathion Doses in California's Medfly Eradication Campaign Using a Physiologically Based Pharmacokinetic Model 189
    Michael H. Dong, William M. Draper, Paul J. Papanek, Jr., John H. Ross, Kimberley A. Woloshin, and Robert D. Stephens

15. An Epidemiological Assessment of the Cantara Metam Sodium Spill: Acute Health Effects and Methyl Isothiocyanate Exposure 209
    Richard A. Kreutzer, David J. Hewitt, and William M. Draper

16. Access to Data for Epidemiological Studies 231
    Ralph R. Cook, Sandra L. Tirey, Nanette W. Spadacene, and Mary A. Woodbury

17. The Successes and Failures of Environmental Epidemiology 245
    Raymond Richard Neutra

INDEXES

Author Index 253

Affiliation Index 253

Subject Index 253

PREFACE

ANALYTICAL CHEMISTS are increasingly able to measure small quantities of chemicals in the human environment. In particular, the analysis of synthetic chemicals (including solvents, chemical intermediates, metals and organometallics, pesticides and formulation contaminants, and combustion products) has advanced to the point at which traces can be detected almost anywhere. As the detection limits have dropped, certain synthetic compounds have been found to be ubiquitous, simultaneously measurable in our food, water, air, and even in our tissues. Although these technological advances have provided a more complete picture of the trace chemical composition of the environment, our understanding of the significance of these residues has not improved. The laboratory animal toxicity of many of the substances, usually administered at high doses, suggests that the exposures detected may also be harmful to humans. To advance the state of knowledge, however, we need some way to test these hypotheses.

Interpretation of the health effects of chemical residues involves health risk assessment, the most sought-after component of which is evidence from well-designed human epidemiological studies. Epidemiology is a health science that examines the distribution of disease in populations and the factors associated with that distribution.

The Environmental Chemistry Division hosted a pedagogical symposium in environmental epidemiology at the Spring 1992 meeting of the American Chemical Society in San Francisco. With support from the U.S. Department of Health and Human Services and the American Chemical Society, a symposium was developed that explored the basic principles and methods of environmental epidemiology. The speakers represented a cross section of experts from academia, the chemical industry, and government. They brought to the symposium a wealth of practical experience in conducting and interpreting epidemiological studies. The objective of the program was to provide an introduction and overview with an emphasis that would be useful to chemists and others interested in environmental chemicals.

This volume, derived from the papers presented at the pedagogical symposium, has been arranged in five sections. The first chapter serves as an introduction. Chapters 2–4, which cover the science of epidemiology, present basic principles of epidemiology, with attention to nomenclature and the design and interpretation of epidemiological studies. Toxicology is inextricably linked with epidemiology in the risk-assessment process, and therefore Chapters 5–8 provide a brief overview of toxicology and toxicological risk assessment. Chapters 9–15 give an in-depth discussion of environmental epidemiology in practice and supply a glimpse of the application of epidemiological methods in human studies. Exposure classification is a major challenge in epidemiological research, and misclassification of exposure groups is a recurring problem. This situation is reflected in many of the chapters. Notably, exposure assessment is also the area in which chemists and biochemists have the most to offer in epidemiological research. The final chapters contributed by two experienced medical epidemiologists examine the unique constraints of epidemiology, an observational science that must bridge numerous legal, ethical, and logistical obstacles, in addition to the more familiar hurdles common throughout scientific research.

This volume will be useful to scientists and engineers, policy makers and risk managers, and others who must deal with public health issues surrounding synthetic chemicals in the environment. The information may be relevant where other types of technology are concerned as well (e.g., nuclear power and radioactive wastes or electromagnetic radiation) as the principles are universal.

Acknowledgments

I thank the symposium speakers, authors, and reviewers who generously took time from their busy schedules to contribute to this unconventional undertaking. The contributions and encouragement of Dean Adams, John Bailar, Margaret Brown, Joy Carlson, Henry Falk, Wolfgang Fuhs, Lynn Goldman, Vernon Houk, Harold Humphrey, Bill Jones, Jack Mandel, Herbert Needleman, Raymond Neutra, Edo Pellizzari, Gerhard Raabe, Cheryl Shanks, Patricia Spear, Colleen Stamm, Robert Stephens, and Warren Winkelstein, Jr., are greatly appreciated. Support for the symposium was provided by Conference Grant R13 CCR907393 from the Centers for Disease Control and Prevention and by the American Chemical Society Committee on Science, Pedagogical Symposium Subcommittee.

WILLIAM M. DRAPER
California Department of Health Services
Berkeley, CA 94704

June 14, 1993

1

Assessing Environmental Risk—Scientifically Defensible or Fantasy?

Vernon N. Houk

National Center for Environmental Health and Injury Control, Centers for Disease Control, Public Health Service, U.S. Department of Health and Human Services, Atlanta, GA 30333

> *Estimating the risk to human health from synthetic toxic substances is becoming increasingly critical in our society. Epidemiological studies should play a vital role in risk assessment. Without human data based on valid, well-done epidemiological studies, extrapolation from animal studies may seriously overestimate or underestimate the risk. To assess risk to the best of our ability, scientists must use all the data—human and animal—and combine this information with the soundest professional judgment. The multistage linearized model for quantitative risk assessment is not appropriate for all chemicals and just because the results of an epidemiological study have been published, they do not necessarily provide the final answers.*

ESTIMATING THE RISK TO HUMAN HEALTH from synthetic toxic substances is increasingly critical in our society. The term "toxic substances" may be a misnomer. These substances have certain toxic effects on animals and humans exposed to large amounts. However, for many if not most of these substances, the effects of low-level, chronic exposure remain unknown or may be only biological, such as induction of hepatic enzyme systems, which for drugs is an acceptable effect but for environmental pollutants is unacceptable.

The main thesis of this chapter is that valid, well-done epidemiological studies of humans should play a vital role in risk assessment. Relying only on extrapolation from animal studies may lead to seriously overestimating or underestimating the risk.

Regulations based on overestimates can have serious, unnecessary economic consequences both by keeping economically desirable products from

This chapter not subject to U.S. copyright
Published 1994 American Chemical Society

being marketed and used and by causing expensive cleanup actions that have no benefit to health. Nevertheless, a completely laissez-faire policy about these substances can be disastrous; we have only to look to Eastern Europe and to some developing countries to see its devastating result.

Over the past decade, scientists have expended many millions of public and private research dollars to investigate and understand the complex relationships between human health and exposure to environmental pollutants. The basic tools of this ongoing search are laboratory studies in animals, studies of molecular biology, and epidemiological studies of humans. Properly done, animal experiments and epidemiological studies provide a basis for linking various human health risks and environmental factors; from these associations, public health and environmental protection policies can be developed to minimize the risks to current and future populations. However, information obtained by means of these tools—animal toxicological studies and human epidemiological investigations—has often been misapplied because of the scientific community's failure to clarify the nature and limitations of our knowledge about environment-related health risks.

Most illnesses clearly caused by chemicals are encountered as exposures in occupational and nonoccupational settings. A number of diseases are known to be caused by exposure to certain chemical or physical agents. Some diseases have no other known cause; these include asbestosis, radiation sickness, caisson disease (decompression illness), and mesothelioma, which is usually caused by asbestos.

In the case of some other illnesses, the chemical–disease link is strong but not unique. Vinyl chloride causes a rare cancer of the liver, angiosarcoma, but this outcome may also be caused by certain arsenicals and androgenic anabolic steroids. The skin disease chloracne is caused by a number of halogenated aromatic hydrocarbons, such as the chlorinated naphthalenes, chlorinated biphenyls, chlorinated dibenzodioxins, some chlorinated azobenzenes, and chlorinated dibenzofurans.

For other diseases, it is even more difficult to establish an actual cause. For instance, benzene has been shown to be associated with a higher incidence of aplastic anemia and myelogenous leukemia in workers who have been exposed to high concentrations of the solvent. Because both aplastic anemia and myelogenous leukemia are also relatively prevalent in the general population, it is difficult to determine in an individual whether the disease was caused by the specific agent of concern or by some other unknown factor.

In addition to cancer, many acute and chronic diseases with potential or perceived chemical causes occur relatively frequently in the general population. These include heart disease and stroke in conjunction with arteriosclerosis, diabetes, chronic obstructive lung disease, arthritis, and immunological and neuromuscular disorders. Other concerns are for congenital malformations and other untoward outcomes of pregnancy. Furthermore, emotional problems, infertility, and psychological disorders among both sexes are often reported by people who fear that their health has been damaged by exposure to chemicals. For all of these conditions, it is rarely possible to demonstrate conclusively a causal role for chemical exposure.

For these reasons, we must use some method of comparison, such as relative risk for exposed and nonexposed populations, and we must be scrupulous in identifying and accounting for all potential confounding factors. We must also be careful to define, quantify, and validate that exposure. When exposure can be validated with laboratory measurements of body burden, in many instances we find that using environmental data and questionnaires results in up to 40% misclassification—about the same as obtained by the toss of a coin. If exposure cannot be validated, we must be careful in interpreting the data, even to the point of admitting that misclassification precludes valid conclusions. In quantitative risk assessment, investigators use the results of high-dose feeding studies and extrapolate the results to untested low-dose exposure levels in the same species; then they extrapolate these extrapolated findings across species to humans.

The factors involved in quantitative risk assessment are scientific fact, consensus, assumption, and science policy. By science policy, I mean the agency's decision about how to handle controversial data. The most certain factor, scientific fact, is usually the least available. We have yet to determine how far the others deviate from the truth. The degree of certainty appears to decrease as one reads through the list.

In "chronic" feeding studies of laboratory animals at the maximum tolerated dose, more than one-half of the tested chemicals have been shown to increase the incidence of tumors. As a result, these chemicals have been classified as animal carcinogens and, by implication, possible human carcinogens, even though (1) many of them have shown little or no mutagenicity, and (2) evidence of human carcinogenesis is lacking. It does not seem to matter, for example, whether the development of these tumors is relevant to human metabolism or even whether the tumors may occur in organs or tissues not found in humans.

During the 1970s, a model was developed by consensus for the carcinogenicity of chemicals. It was based on experience with radiation, which defined a linear relationship between dose and response over a wide range of exposures. Study results showed that ionizing radiation produced genetic mutations that led to tumor development in exposed populations. These radiation-induced mutations were observed in animals, plants, and bacteria. By using bacteria, investigators could study the response curve produced by very low doses.

This radiation experience of seemingly unending cellular response to ever-decreasing radiation doses stood in stark contrast to a fundamental rule of chemicals in toxicology—that is, the dose makes the poison. Thus, consensus abrogated a long-held principle of toxicology and, with scant evidence, determined that for chemically induced carcinogenesis, no exposure is free of threat.

The radiation model for carcinogenicity was based on the simple concept that all cancer is caused by mutation of the cellular DNA. Because results of animal studies had confirmed a dose–response relationship for radiation and tumor development, long-term, high-dose bioassays became the choice for studying the potential carcinogenic effects of chemicals. A dose approaching the maximum tolerated dose was selected to ensure that no positive response was missed and because the fewer animals needed to demonstrate a response meant a less expensive test.

For the past 20 years, rodent feeding has played the leading role in determining the carcinogenicity of chemicals. Regulatory science has largely lost sight of the basis of the radiation model that required not only a dose-dependent response but, more important, agent-caused genetic mutations. We have been driven by the end point, cancer development, forgetting that we must also understand the means to the end. Thus, a host of chemicals have shown an increased tumor incidence, compared with controls. Most show little or no mutagenic activity. Over the last two decades, the purpose of lifetime bioassays has shifted from investigating the mechanism of carcinogenicity to accumulating data from which to calculate the supposed human cancer potential of chemicals.

Many scientists now understand that loading an animal with a chemical for a lifetime for the purpose of counting tumors in order to satisfy a mathematical extrapolation model, does not necessarily predict its potential for carcinogenicity in humans. In the early days of risk assessment, this modeling approach was the "only game in town". It combined some animal data, statistics, and mathematical extrapolation to evaluate which chemicals may produce a specific human health effect. This combination process, in its many forms, became the basis of science policy.

Most scientists now recognize that not all chemicals fit the radiation model of carcinogenesis. Thus, for nongenotoxic chemicals, we need another approach that allows us to properly protect the public's health without wasting resources because of excessive regulation, as may be dictated by the linearized multistage extrapolation model.

Although the results of animal studies may imply carcinogenicity in humans, the results of epidemiological studies have identified chemicals that have proved to be human carcinogens. From the association of scrotal cancer in chimney sweeps with soot to liver angiosarcoma of plant workers with vinyl chloride—to name only two examples—epidemiology has identified the relationship. We must not disregard the results of human experience when evaluating the implications of animal studies. The evidence gained from human studies is strengthened by consistency among several studies. Conflicting results among well-done, large epidemiological studies raise serious doubts about apparent associations.

It becomes evident that many factors influence the development of disease. Some illnesses, including many cancers, may have latency periods of

20–40 years. Moreover, in general, environmental exposure to synthetic chemicals has been at relatively low concentrations and through a variety of routes—inhalation, ingestion, and absorption through the skin. As a result, it is now impossible to determine precisely for the individual the events that led to the development of disease. The state of the art in medical science or epidemiology is not such that we can predict with certainty whether a person who has been exposed to chemicals will ultimately develop a particular disease or condition.

In most cases, the conclusion must be drawn that the scientific database now available does not permit a certain determination of whether exposure has a causal relation to illness in humans. The data now available do provide sufficient evidence to reduce exposure, and thus possibly to prevent disease in the future. There is a reason for this dichotomy between prevention and attribution of cause. The studies that generate information about the chronic low-dose toxic effects of chemicals do not permit predictions with full confidence about the health of an individual, but they do assess the health of a population and what degree of risk a given population will run if exposure continues.

Proper use of the scientific data can lead to major public health benefits; the application to that purpose is both responsible and just. However, to press such data into service to explain the cause of an individual's disease carries a great potential for misuse of the data.

Epidemiological studies never prove cause and effect, although in some instances, reasonable people would accept them as proof. It is not ethical to purposely expose individuals to hazardous substances. In studies of humans, investigators must find only instances of inadvertent exposure, and we must design studies that provide the best-possible evidence for or against an association.

To assess risk optimally, we must gather all the data—human and animal—and combine them with the soundest professional judgment. We must recognize that the multistage linearized model for quantitative risk assessment is not appropriate for all chemicals. At the same time, we must acknowledge that the results of a published epidemiological study are not necessarily the final answer. Many of the epidemiological tools, unless used impeccably with large enough populations, will yield inconclusive results—neither positive nor negative.

Our major limitation is dose quantification, a problem that is relatively simple to solve for laboratory animals. In epidemiological studies, investigators frequently use qualitative indicators of dose, such as "high", "medium", and "low" or "yes" and "no". As stated, such studies have a significant problem with misclassification among the potentially exposed and unexposed groups.

To illustrate the need to base public health judgment on all data combined, two examples are provided.

Lead

Early on, adverse consequences of exposure to lead were associated with the workplace, affecting the workers and their families and those in the immediate vicinity of the workplace through environmental releases. Others have been exposed through ingesting material that contains lead. These are primarily young children who ingest paint chips and soil and dust that contains very small particles of lead. The widespread use of leaded gasoline as an automobile fuel introduced another source of lead, and inhalation became a significant route of exposure in high-traffic areas, not just in the vicinity of smelters.

Laboratory animal data for lead provided both a lowest observed adverse effect level (LOAEL) and a no observed adverse effect level (NOAEL) for exposure to lead. These early animal values were, for some end points, greater than or equal to levels seen in workers and their children. For children in the United States before the mid-1960s, a level of lead below 60 µg/dL in whole blood was not considered dangerous enough to require intervention. Subsequent research noted adverse health effects on humans with lower blood levels; in 1985 the threshold was lowered to 25 µg/dL, and more recently to as low as 10 µg/dL.

Studies of humans have since demonstrated adverse effects at lower blood lead levels. Effects of in utero exposure include decreased gestational age and birth weight and retarded mental development. The effects in children, for which no threshold has been defined, include decrements in IQ and hearing, diminished growth, and reduced vitamin D metabolism. For adults, they include increased blood pressure in men. Because of the effects of low-dose exposure, the level of blood lead warranting concern in the United States has been reduced to 10 µg/dL. This threshold was selected not because lower levels are without consequences, but because of a practical need to reduce the current blood lead levels in the general population.

Dioxin

For the class of chemicals known as dioxins and furans and specifically for 2,3,7,8-tetrachlorodibenzo-*p*-dioxin, we now have more scientific facts than for most chemicals. For more than a decade, the results of animal studies have shown dioxin to be the most potent carcinogenic synthetic chemical tested. It has also been said that it is "the most toxic man-made chemical known to man". We now recognize that most, if not all, of the biological and toxic effects of dioxin are receptor-mediated. We know that in animals each of these cellular manifestations of dioxin-receptor-mediated activity shows a concentration below which there is no observable cellular response.

Dioxin is not considered to be genotoxic. If its mode of action for carcinogenesis is either receptor-mediated or toxicity-induced cell proliferation, the available scientific information indicates that the linearized multistage model is inappropriate for estimating excess lifetime cancer risk to humans from the results of animal studies.

Before we could measure dioxin in humans, the most consistent physical marker of high-level human exposure to dioxin was chloracne. More than 450 cases of chloracne have been recorded in workers involved in eight "accidents" in trichlorophenol plants that occurred between 1949 and 1968 in several countries around the world. Other cases have occurred in hundreds of workers exposed to routine leaks and spills. Heavy exposures, usually stemming from production "accidents" that contaminate workers with (2,4,5-trichlorophenoxy)acetic acid, dioxin, and other chemicals have been related to transient neurological and liver effects, which in almost all cases disappeared with the passage of time.

One of the most heavily exposed populations in the world is that around Seveso, Italy. That population of about 30,000 individuals under study includes about 700 who lived in or who were in the most contaminated area. We have measured dioxin content of up to 56,000 parts per trillion in the lipid portion of serum in some of those individuals or about 10,000 times the mean background levels for the U.S. population. Chloracne is the only disease yet found in the Seveso population. The mean dose to the Seveso children who developed chloracne, about 3 µg/kg, is three times as high as the dose of 1 µg/kg that kills half of the guinea pigs exposed to it.

The recently completed Centers for Disease Control National Institute for Occupational Safety and Health (NIOSH) study of cancer mortality in workers exposed to dioxin adds considerable but not conclusive evidence about its relationship to cancer. Exposure in this population has been validated by laboratory measurements of dioxins. This is a study of more than 5000 production workers in the U.S. chemical industry and is probably the largest and most elegant study that is possible. In general, the study showed that over 1500 individuals who had the highest exposure (600 times background) had a modest increase of all cancers combined [standard mortality rate (SMR) = 146] and of cancers of the larynx, bronchus, and lung (SMR = 142). There were no increases in cancer in the 1500 individuals who had been exposed for less than 1 year but whose serum dioxin levels were 60 times background levels, and there was no increase in the 2000 other individuals who had lesser exposures. In addition, the contribution of exposure to other chemicals confounding the results in this NIOSH study cannot be fully evaluated.

With the exception of chloracne—a potentially disfiguring but non–life-threatening skin condition—there are no convincing data that link exposure of humans to dioxin, even at very high levels, with early mortality, adverse reproductive outcomes, or chronic diseases of the liver or the immune, cardiovascular, or neurological systems. Although the question of its link with cancer is not settled, if dioxin is a human carcinogen, I believe it is a weak one that is associated only with very-high-dose exposures. Furthermore, I believe that dioxin, if it is a human carcinogen, acts as a promoter, not as an initiator. Two well-done studies in workers support the last two conclusions.

Conclusion

Dioxin and lead are good examples of why we must use both animal and human data to evaluate the potential health effects of chemical exposure in humans. The early animal studies of lead seemed to show no effects at concentrations greater than those experienced by humans from many environmental exposures. Further investigations with humans, however, have shown several effects for which no threshold has been defined. Thus, we continue efforts to remove lead from use and to reduce its environmental release to the greatest extent possible.

We now have good information about dioxin's mechanisms of toxicity in animals. The no-effect level for various toxicological effects in animals is a daily dose of about 1000 pg/kg. By using kinetic data, we can equate this to a daily dose of 100 pg/kg body weight in humans. We also have good information on the effects on human health under steady-state conditions. These data indicate that the environmental levels of dioxin to which the general population is now exposed are not enough to warrant concern for human health. We do not want that exposure to increase, but neither need we undertake expensive environmental cleanup actions where the contamination has already occurred.

Combining all the data about health risk for a particular substance may not provide the conclusive answers that are so much in demand. Nevertheless, we believe that we can demonstrate strong associations, where they exist, between exposure and adverse health outcomes so that reasonable people can take reasonable actions to protect public health and the environment. Thus, we see the role of public health agencies as primarily one of prevention—just as it has always been.

RECEIVED for review September 3, 1992. ACCEPTED revised manuscript April 3, 1993.

The Epidemiological Method

Gary H. Spivey

Department of Epidemiology and Industrial Hygiene, Unocal Corporation, 1201 West 5th Street, Los Angeles, CA 90017

> *Epidemiology is the study of the distribution of disease in populations and the reasons for that distribution. Disease distribution can be characterized by person, place, and time. Risk factors increase the probability of disease. Epidemiologists identify associations between potential risk factors and disease and evaluate those associations for evidence of causality. Latency, the time between exposure and onset of disease, must also be considered in assessing causality. Associations between risk factors and disease are shown by increased risk relative to those without the risk factor (relative risk) or by proportion of disease associated with a given risk factor (attributable risk). Frequency of disease is expressed as incidence (new cases in a time period) or prevalence (cases existing at a point in time).*

EPIDEMIOLOGY IS A HEALTH SCIENCE that studies the distribution of disease in populations and the factors associated with that distribution. As the definition implies, epidemiology focuses on groups of people, not on individuals. Because epidemiologists try to find the causes of disease, they have earned the nickname "medical detectives".

History

Throughout history, much of human activity has been determined by the impact of disease. For example, the defeat of Athens by Sparta was in large part due to an epidemic of what was probably scarlet fever (1). The great plagues of the Middle Ages and the more common lesser epidemics created great fear. This in turn led to the development of the first death certificates, called "bills of mortality". These documents were instituted to provide the upper class with advance warning of developing epidemics so that they could leave the cities (2).

In 1662, John Graunt published the first systematic study of these bills of mortality (3). He discovered many of the basic facts of the typical patterns of mortality. Further development of vital statistics and statistical methods came in the 19th century with the establishment of national censuses, the desire to better understand means of preventing epidemics, and the desire to evaluate the efficacy of medical interventions (4).

During the same time period, methods of careful field observation of the progress of epidemics in populations were developed. This advance was aided by the development of geographical and historical pathology, which focused on characteristics of people, places, and time in relation to disease occurrence (5). By careful epidemiological observation and reasoning, a number of major advances in disease prevention were made before Pasteur and the discovery of microorganisms. Examples included control of childbed fever, cholera, and typhoid (6).

During the 1940s and 1950s, epidemiology began to shift from the study of infectious diseases to that of chronic diseases, such as heart disease and cancer. During that time, it also became recognized as a scientific discipline in its own right. Rapid advances in methodology were made during the 1960s and 1970s. The subspecialty of environmental epidemiology began to emerge in the middle to late 1970s.

Epidemiological Observation

One of the earliest methods, and still the heart of epidemiology, is careful epidemiological observation. The most fundamental tenet of epidemiology is that disease does not occur randomly. Therefore, if you can measure and describe where, when, and in whom disease does occur and what other entities follow the same pattern, you can discover why the disease occurs. The basic method of describing disease distribution is description of the characteristics by person, place, and time.

"Person" refers to the personal characteristics of those who develop disease. The most important of these is age. All diseases vary sharply with age. Other important personal characteristics include sex, race, education, occupation, and personal habits, such as cigarette smoking.

"Place" refers to geographical variation on both macro- and microscales. Differences in disease occur among continents, within countries, and across cities. On a microscale, one may link cases of a disease to a single household or restaurant, for example.

"Time" refers to characteristics of when the disease occurs. There are long-term trends in many diseases, such as the rise during the past 20 years of lung cancer among women because of changing smoking habits (7). There are also cyclic trends in many diseases. Influenza, for example, is more common during the winter.

One characteristic of time important to understanding infectious disease is the incubation period (Figure 1). This is the time between exposure and onset of clinical symptoms. For chronic diseases, we refer to the period between exposure and onset of disease as "latency". For most chronic diseases latency is measured in decades. This long latency period is one of the major obstacles to environmental epidemiology, because it is difficult to assess exposures that took place decades earlier.

The person–place–time method can be illustrated by an example from 16th-century England. Sir George Baker became interested in a strange disease noted in Devonshire, called the "endemial colic of Devonshire" (8).

Baker first noted that the disease occurred only in Devonshire (place), as implied by the name. This alone suggests the possibility of an environmental or life-style cause. He next noted that the disease occurred only in autumn (time). Then he noted that the disease was far more common among the lower classes (person). This simple observation of distribution by person, place, and time led to further investigation of what happens in autumn in Devonshire. He found that autumn is the time of the apple harvest and of cider pressing. Looking more closely at place, he found that the greatest number of cases were seen in the parts of Devonshire where cider was made. He also observed that the lower classes drank a great deal of cider, whereas the upper class did not. Thus, the disease appeared to be related to the pressing of cider. However, other parts of England pressed cider and did not have this disease. The next step was to compare the equipment and methods of Devonshire with those of other parts of England. In Devonshire, the stone cider presses were sealed with lead, whereas other parts of England did not use lead: The endemial colic was what we now recognize as lead poisoning.

Infectious Diseases

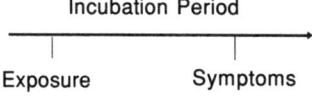

Chronic Diseases

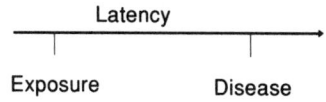

Figure 1. Time-line illustration of incubation period and latency.

Today, we seldom have circumstances as clear-cut as in this example. Person, place, and time characteristics are typically delineated by means of a series of investigations. When we conduct studies, we seek to identify characteristics that vary in frequency in the same pattern as the disease; that is, when the disease is rare, the characteristic is rare, and vice versa. For example, early studies of acquired immunodeficiency syndrome (AIDS) in the United States showed that AIDS was more frequent in populations with a higher proportion of homosexuals or intravenous drug abusers (9). In Africa, AIDS is more common along major truck routes (10). Such characteristics are said to be *associated* with the disease. The association between characteristic and disease must be assessed to determine whether the relationship may be causal; that is, whether the characteristic causes the disease or whether the relationship has some other explanation, such as error or a close association with another characteristic.

Characteristics that are judged to be important in determining when and where disease occurs are called "risk factors". Risk factors may not be the root cause of the disease, but having a risk factor increases one's chance of developing the disease. In the example of the endemial colic, identified risk factors included living in Devonshire, being of the lower class, and drinking cider.

Measuring Disease

Although we frequently treat disease as if it were a yes-or-no phenomenon—either you have it or you do not—it is much more complex than that. All disease has some variability in its manifestation; frequently, some cases go completely unnoticed. The variation can be extremely small; rabies, for example, is virtually 100% fatal. The main variation is in the length of course of the disease. On the other end of the spectrum is disease such as mumps, for which 30% of children develop immunity without ever showing clinical signs of illness, although 1 child in 5000 dies of complications of the disease. We call this range of clinical manifestations the "spectrum of disease" (Figure 2). It can complicate the study of disease because of the complexity of de-

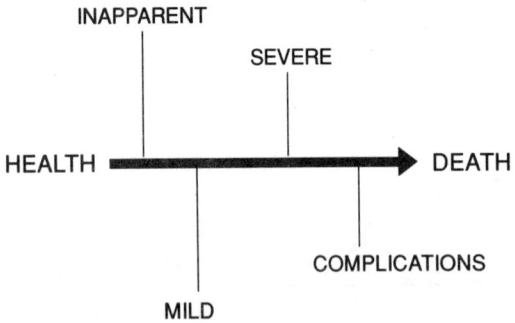

Figure 2. The spectrum of disease.

termining who does or does not have the disease. Sometimes we can determine at least some risk factors, such as malnutrition or genetic predisposition, that influence the severity of the disease process. In most cases, we do not know these factors and simply refer to "biological variability."

Besides information on the definition and diagnosis of disease, measurement of the frequency of the disease requires knowledge of the population size. Suppose one were interested in the frequency of heart attacks in California. If it were found that there are 20,000 cases per year in California and 100,000 cases per year in the rest of the United States, could one conclude that heart attacks in California are much less common than in the rest of the United States? Clearly, the answer is "No". If the population of California were 30 million, the frequency of heart attacks would be 20,000 in 30 million, or 1 in 1500 people. If the population of the rest of the United States were 200 million, the frequency of heart attacks would be 100,000/200 million, or 1 in 2000 people. Thus, the actual frequency in California would be higher than in the rest of the United States. (Of course, these figures are made up.)

We need to be a bit more precise. Typically, we measure frequency of disease by one of two definitions: incidence or prevalence. Incidence is the number of new cases divided by the size of the population. It is a measure of the "speed" at which the disease is occurring. Just as we must know the length of the observation period in calculating the speed of a car, so we must specify the time period in calculating incidence. For many diseases, a large proportion of the population may not be susceptible. For example, anyone who had measles as a child or was vaccinated is typically no longer at risk of developing the disease. For such diseases, relating the number of new cases to the total population would give an inaccurate measure of the disease's frequency. What we are really interested in is the number of cases among those actually at risk. Thus, the full technical definition of incidence is the number of new cases in a time period, divided by the number of people at risk. This produces an estimate of the chances of a susceptible person getting the disease.

$$\text{incidence} = \frac{\text{number of new cases (time period)}}{\text{population at risk}}$$

Prevalence is the number of *existing* cases of disease at any given *point in time*. In this case, we do not correct for those at risk. Prevalence is simply the number of existing cases at a point in time, divided by the total size of the population.

$$\text{prevalence} = \frac{\text{number of existing cases (point in time)}}{\text{total population}}$$

In Figure 3, prevalence is represented as liquid in a pot. The greater the flow of water into the pot, the higher the water level tends to rise. Likewise,

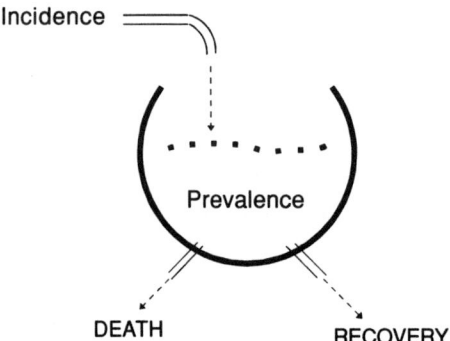

Figure 3. The "prevalence pot."

the greater the flow of water out of the pot through the drains, the more the water level will fall. The flow into the pot represents incidence. The flow out represents the two ways in which a case can drop from the pool of prevalent cases: recovery from the disease and death. A disease with a very short course, for example, 1 day, can have a low prevalence, even if it occurs commonly, because of the high outflow through recovery or death. A disease of long duration, for example, 20 years, can have a high prevalence, even if it is relatively rare, because of the low outflow. Prevalence and incidence tend to be in equilibrium, with prevalence being proportional to incidence times the length of the disease (duration).

$$\text{prevalence} \propto \text{incidence} \times \text{duration}$$

This relationship is significant because prevalence rates are frequently compared, being easier to obtain than incidence rates. When we compare prevalence rates, we assume that differences in prevalence reflect differences in incidence. A population that has a higher prevalence of a disease may, in fact, be at higher risk of the disease. However, that population could have a lower risk but a longer duration of disease, for example, because of better medical care. Because of this problem, incidence is the preferred measure for studies of causation. Prevalence can be useful for public health purposes, such as estimating the need for nursing home beds.

Measures of Association

Just as we have measures of frequency, we also have measures of association. Most commonly, we use relative risk and attributable risk.

Relative risk (*RR*) is a measure of how many times greater is the risk of one population than another. It is defined as the incidence among those exposed to a risk factor (I_e), divided by the incidence among those not exposed (I_0).

$$RR = \frac{I_e}{I_0}$$

Relative risk is a measure of the strength of an association. The greater the relative risk, the more likely that the risk factor is important in causation. Generally, we tend to doubt relative risks of less than 1.5. Lower relative risks are more likely to be artifacts due to uncontrolled confounders. They suggest that the factor may not be a very important cause of the disease. Somewhat like a low chemical detection limit, where the presence of the chemical at that level does not necessarily indicate harm, a relative risk less than 1.5 typically indicates an association that may be due to confounding or that is not biologically important.

Attributable risk is a measure of the impact of a risk factor. We may be interested, for example, in the impact of cigarette smoking on lung cancer. Not all lung cancer is caused by cigarettes. There is a normal background rate of lung cancer in nonsmokers. Even among smokers, some of the lung cancer can be accounted for by the normal background risk. The attributable risk (AR) then is defined as the difference between the incidence in the exposed and the unexposed—in other words, the excess incidence among the exposed after removing the expected background incidence.

$$AR = I_e - I_0$$

Attributable risk is the portion of the incidence among those exposed that can be attributed to that exposure. It is frequently expressed as a percentage of total incidence (Figure 4). Like prevalence, it is more useful for public health purposes than for etiological research. As an illustration (Table I), cigarette smoking's strong association with lung cancer is shown by a very

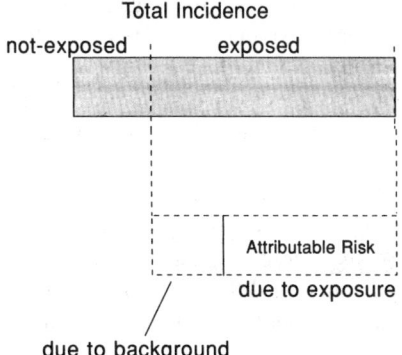

Figure 4.   Illustration of attributable risk.

Table I. Measures of Risk: Relative versus Attributable

Factor	Lung Cancer	Coronary Heart Disease
Heavy smokers (incidence)	166/100,000	599/100,000
Non smokers (incidence)	7/100,000	422/100,000
Relative risk (ratio)	23.7	1.4
Attributable risk (cases/100,000)	159/100,000	177/100,000

SOURCE: Data are taken from reference 11.

high relative risk. The attributable risk is also high. The association of cigarette smoking with heart disease, however, does not appear to be very strong. The relative risk is small, meaning that cigarette smoking is not a major independent cause of heart disease. However, cigarette smoking has a large attributable risk with heart disease, and thus, has major public health significance.

Measuring Exposure

Because exposure is the major independent variable, the value of an environmental epidemiological study depends in large part on the quality and accuracy of the exposure estimate. Several general issues related to measuring or estimating exposure must be considered.

The technical measurement of exposure may be difficult because of such problems as imperfect knowledge of environmental behavior or the expense of monitoring methods. It is further complicated by the fact that there are several routes of entry into the human body, particularly inhalation, ingestion, and skin absorption.

There also may be multiple sources of exposure. For example, one can be exposed to benzene while pumping gas in a service station or while being in the same room with a cigarette smoker. The importance of different sources and routes of exposure may be affected by each individual's life style, such as the amount of time spent outdoors and the activities engaged in.

As mentioned earlier, the exposure of interest may have been in the past. Because of trends in industrial activity, transportation, and regulation, today's exposure may not accurately reflect past exposure. For past exposure, the most we can hope to achieve is a relative gradient of exposure, such as high, medium, or low.

Finally, because epidemiological studies generally involve large numbers of people and long periods of time, we must represent exposure by some kind of summary measure, typically an average value. Very little information is available on the relative importance of different types of summary exposure statistics. However, some data on acute effects of both nitrogen dioxide and ozone suggest that the peak level is more important than the average. In a somewhat different fashion, a severe sunburn early in life may be the most relevant sun exposure for future skin cancer. Thus, we need to realize that alternate ways of summarizing the data may be critical.

Causality

Judging whether an association is causal is a difficult task in epidemiology, because epidemiological studies are observational, not experimental. The investigator has little control over the independent variables. An important first step is to look for possible bias that might have produced a spurious association. Bias is a systematic error in measurement, such as the results of an improperly marked ruler. Bias can arise in many other ways in epidemiological studies, such as by using nonrandom samples of a population. A good example of this type of bias was the poll in the 1936 presidential election that indicated Landon had defeated Roosevelt. The poll was based on the telephone directory. The pollsters failed to recognize that in 1936 only the more affluent members of the population generally had telephones, and those people were more likely to vote Republican (12).

We now consider causal evidence (Figure 5). In the first model, A causes X. In this case, A is said to be both necessary for X to occur and sufficient for X to occur. In the second model, A is necessary for X to occur but is not sufficient because B is also necessary. The tubercle bacillus is necessary for tuberculosis to occur, but it is not sufficient. Not everyone infected with the tubercle bacillus will develop tuberculosis. In the third model, A is sufficient for X to occur but is not necessary because there are other causes. Cyanide is sufficient to cause death, but there are many other causes. The fourth model is by far the most common. In this case, A is neither sufficient nor necessary for X to occur.

Because of most causal factors being neither necessary nor sufficient, nonexperimental conditions, and many potential sources of bias, the assess-

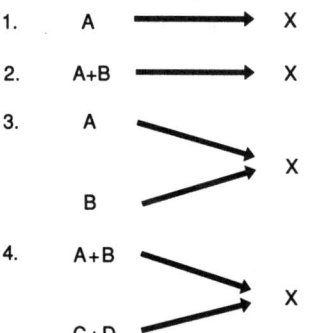

Figure 5. Models of causality.

ment of causality must be done with care. Epidemiologists generally use a common set of criteria to help judge the evidence. They are the following:

- time sequence
- strength of association
- specificity of association
- consistency among studies
- consonance with existing knowledge

The first criterion is time sequence. To be a cause, the factor must precede the illness. Although the criterion is conceptually simple, actually verifying this sequence can be difficult.

The second criterion has already been discussed: the strength of the association. This is usually measured by the relative risk.

The third criterion is specificity. The easiest is a one-to-one association—one factor, one disease—but this situation is rare. Generally, the fewer diseases a factor is associated with, the more likely it is thought to be causal. However, many diseases are known to be caused, in part, by cigarette smoking. These include multiple types of cancer and such diverse other diseases as aortic aneurysm, ulcers, and fetal death. Thus, cigarette smoking is one causal agent that does not meet the specificity criterion. Of course, one can also say that cigarette smoke is not really a single agent because of its complex mix of chemicals.

The fourth criterion is the consistency of the association. Can the findings be replicated in different studies and in different populations? If there is a true biological relationship, it should be found among different ethnic and racial groups, both sexes, and so on; or there should be a good reason why the relationship does not hold for a particular group. For cigarette smoking, the association with lung cancer has been shown by many different studies, by means of different methodologies, and in many different populations all over the world.

Finally, is the finding consonant with existing knowledge? Do all the bits of evidence constitute a coherent whole? In particular, we look at whether the proposed association is biologically plausible. If not, it may be because of something we do not understand about biological systems, but it is more likely to be a noncausal association.

References

1. Cartwright, F. F. *Disease and History;* Rupert Hart-Davis: London, 1972.
2. Susser, M. *Causal Thinking in the Health Sciences, Concepts and Strategies of Epidemiology;* Oxford University: New York, 1973; p 17.
3. MacMahon, B.; Pugh, T. F. *Epidemiology, Principles and Methods;* Brown and Company: Boston, MA, 1970; p 6.
4. Lilienfeld, A. M. *Foundations of Epidemiology;* Oxford University: New York, 1976; p 23.

5. Gordon, J. E. In *Epidemiology of Health;* Galdston, I., Ed.; New York Academy of Medicine, Health Education Council: New York, 1953; p 44.
6. Gordon, J. E. In *Epidemiology of Health;* Galdston, I., Ed.; New York Academy of Medicine, Health Education Council: New York, 1953; pp 52–53.
7. "Smoking and Health"; report of the Surgeon General, U.S. Department of Health, Education, and Welfare. U.S. Government Printing Office: Washington, DC, 1979, p 5–17.
8. Baker, G. *An Essay Concerning the Cause of the Endemial Colic of Devonshire;* J. Hughs: London, 1768. Reprinted by Delta Omega Society, 1958.
9. Quinn, T. C. *Bull. Pan. Am. Health Organ.* **1989**, *23:(1–2)*, 9–19.
10. Orubuloye, I. O. *Int. Conf. AIDS* **1992**, *8(2)*, D425.
11. Mausner, J. S.; Bahn, A. K. *Epidemiology, An Introductory Text;* W.B. Saunders: Philadelphia, PA, 1974; p 322.
12. Orubuloye, I. O. *Int. Conf. AIDS* **1992**, *8(2)*, 322.

RECEIVED for review September 3, 1992. ACCEPTED revised manuscript February 17, 1993.

3

Design Methods for Occupational and Environmental Epidemiology

MaryFran Sowers

University of Michigan, School of Public Health, Department of Epidemiology, Ann Arbor, MI 48109

> *This chapter provides an introduction to the fundamental epidemiological study designs, including the cross-sectional, cohort, and case-control designs. Each has strengths or weaknesses relative to the ability to characterize a temporal relationship of exposure to disease. The discussion of each design includes a description and example of its application or a potential limitation.*

THE CONCEPT OF RISK in epidemiology (the probability of disease occurrence) is used to identify factors that cause or are highly associated with disease. These risk factors can be individual attributes, specific exposures, or individual behaviors that alter, positively or negatively, an individual's likelihood of having a specific disease.

Epidemiological studies compare groups of individuals having different levels of exposure to determine whether their disease frequencies are different from those of groups having no exposure. In certain epidemiological study designs, one can determine the amount of risk associated with a particular exposure by subtracting the disease frequency in nonexposed (or less-exposed) individuals from that of exposed individuals.

Epidemiological studies can serve a variety of purposes in environmental epidemiology and address a wide spectrum of questions. This chapter reviews the primary study design options available to the scientist for characterizing health risks in populations. It delineates design options, the strengths and weaknesses of each design option, examples of each option, and the type of effort required for implementation. The references include a list of suggested readings to facilitate a more in-depth understanding of the study designs and their quantitative output (1–5).

The epidemiological study design options commonly used in environmental epidemiology studies are shown in the following list. They include descriptive studies, such as cross-sectional or correlational studies. Observational analytic studies include cohort and case-control studies.

Types of Epidemiological Studies

I. Descriptive studies
   A. Cross-sectional studies
   B. Descriptions of clusters
   C. Ecological studies
II. Observational Analytical studies
   A. Cohort studies
      1. Prospective cohort
      2. Retrospective cohort
   B. Case-control studies
   C. Intervention studies

Descriptive Studies

Cross-Sectional or Prevalence Studies. Cross-sectional studies characterize both exposure and disease outcome at the same point in time. A common variation is the prevalence study, which correlates the prevalence of a particular disease (existing cases of disease) with self-reported historical exposure information. The cross-sectional design cannot distinguish whether exposure occurred before or after disease developed. Because its estimates the prevalence of a particular disease, rather than the amount of new disease (incidence) associated with that exposure, the cross-sectional study design is considered a hypothesis-generating, rather than hypothesis-testing, study design. This design is more efficient for describing the disease experience of the population and attendant allocation of health care resources than for determining the role of the exposure in disease causality.

Probably the most notable limitations of the cross-sectional studies are their inability to establish a temporal sequence of events and the potential for a survivor bias. Furthermore, the cross-sectional study is an inefficient tactic if the outcome is a rare condition.

Descriptions of Clusters. "Clusters" are reported aggregations of disease that appear to occur with greater frequency in a particular locale and time frame than would typically be expected. Recently, guidelines have been issued by the Centers for Disease Control for the investigation of clusters (6). In part, these guidelines are an acknowledgment that the *perception* of a disease cluster may be just as important as the cluster.

Ecological (Correlational) Studies. In ecological studies, the units of observation are groups, rather than individuals. These studies are used when exposure and disease data have been collected for separate purposes and it is not possible to link a specific exposure and/or disease to the individual. The overall levels of exposure and disease for the groups are compared to determine whether those with the greater exposure have greater disease frequency.

A famous correlational study yielded findings, later validated by clinical trials, that areas with greater fluoride levels had fewer individuals with decayed, missing, or filled teeth. The initial observation generated testable hypotheses, but the relationships were not considered causal.

The primary limitation of the ecological study is that an association at the group level may not necessarily indicate a causal association between the exposure and outcome at the individual level. This is known as the "ecological fallacy". Whereas residents in an area with an incinerator may have higher lung cancer rates, they also may be more likely to smoke. In an ecological study, it is not possible to determine which exposure is the causal factor. If both were causal, it would not be possible to describe how much of the risk should be attributed to either exposure.

Analytical Observational Studies

Cohort Studies. Cohort studies involve large groups of people who have been considered as to their exposure at a specified base-line time and subsequently observed for the development of disease. The cohort study takes advantage of the incidence measure to estimate risk. If disease is related to exposure, the frequency of exposure should be greater among the diseased than among the nondiseased group. The estimator of the risk represents the probability of developing disease, given a particular exposure. At the time the exposure is defined, the cohort is also examined to determine that it is free of preexisting disease so that a temporal relationship between exposure and disease can be maintained.

This approach is found in occupational studies in which workers in specific industries are characterized as to potential exposure and monitored over the course of time for changes in exposure until disease is detected. Well-known examples of environmental cohort studies include the Atomic Bomb Survivors Cohort of Hiroshima, Japan, and the Ranch Hands Cohort, who were on the flights to apply defoliants in Vietnam.

Variations in the cohort design are based on the *type of group* required to formulate exposed and nonexposed components of the cohort and the *time frame*. The element of time frame is shown in Figure 1.

In a prospective cohort study, persons are classified with respect to exposure at the beginning of the study and followed for a specified time into the future to determine disease outcomes. In contrast, a retrospective (his-

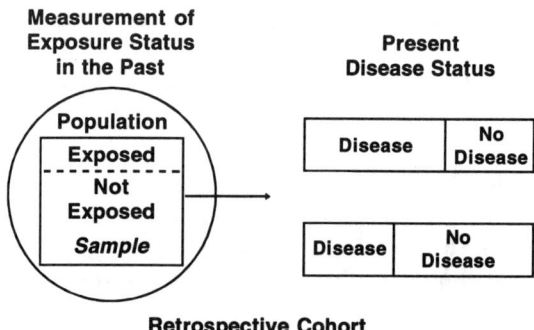

Figure 1. A comparison of the time frames for prospective and retrospective cohort studies.

torical) cohort study utilizes groups of people identified in the past and uses historical records to estimate their exposure. The outcome, either a morbid or a mortal event, is determined in the present.

The selection of individuals to participate in a cohort study is often determined by the type and frequency of exposure being evaluated. For a common exposure such as smoking, it would be relatively easy to formulate a randomly sampled community cohort. For more rare exposures, such as those often associated with occupations or environmental factors in a specific geographical area, it is more efficient to select a group of individuals who have undergone a specific exposure and to subsequently examine the health effects. This kind of selection process has been used in cohort studies of individuals in occupations such as rubber processors, uranium miners, and shipbuilders. It has also been used to study residents in geographical areas with nuclear power plants and hazardous material dump sites. This use of a special-exposure cohort generates enough persons exposed to the factors of interest that reasonably stable estimates of their risk can be formulated. Additionally, special-exposure cohorts permit the evaluation of outcomes sufficiently rare that a random population cohort would be prohibitively large (and therefore costly).

An example of a special-exposure cohort can be found in our studies of fluoride and bone fracture in Iowa. When the cohort was formed in 1983, the initial hypothesis was that fewer fractures would be found among women exposed to higher levels of fluoride than among women exposed to usual levels of fluoride. Higher fluoride levels were observed among women who lived in a community in which the naturally occurring fluoride level is 4 parts per million (ppm), the highest allowable level under current Environmental Protection Agency standards. Usual levels of fluoride were characterized in women who lived in communities in which the water supply was fluoridated to a level of 1 ppm. In contrast to the hypothesized outcome of fewer fractures, we actually found more fractures during the 5-year period of observation. Women with higher fluoride exposure at base line were found to have a two times greater risk of fracture in the subsequent 5-year period than women with the lower fluoride exposure, after adjustments were made for age, body size, perimenopausal estrogen use, and thiazide antihypertensive medication use (7).

The cohort study is a powerful design for defining incidence and investigating potential causes of disease. Because exposure is measured before the presentation with disease, one can maintain an appropriate time sequence that is not reliant on an individual's correct self-reporting of characteristics. This temporal sequence is considered one of the essential criteria for causal inference. Cohort studies are particularly appropriate for studying fatal diseases, for which any recall of exposures must come from a surrogate who may not know personal characteristics of the decedent or whose recall of certain characteristics may be highly influenced by the process of dying.

The cohort study is limited by the sample size required for ascertaining important differences in the distribution of disease. Even with relatively common diseases, such as cardiovascular disease, sample participants usually number in the thousands.

Case-Control Studies. Case-control studies utilize samples of diseased and nondiseased persons to determine the frequency of the exposure of interest, rather than to evaluate exposure in a disease-free population and to await subsequent disease expression as would be done in a cohort study. If exposure is associated with outcome, the frequency or the gradient of the exposure should be greater in the disease group than in the control or nondiseased group. The quantitative estimator of this relationship is called an *odds ratio*, which identifies the probability of odds of exposure among the diseased in comparison to the odds of exposure in the control group and is considered a surrogate estimate of the relative risk (as shown in Figure 2).

The case-control study design is widely used in environmental epidemiology. For example, London et al. (8) published a case-control study of the relationship between exposure to electric and magnetic fields in the home and risk of leukemia among children aged 0–10 years in Los Angeles County, California. The cases were ascertained through a population-based tumor registry. There were two kinds of controls: friends of those studied and a random population control secured through random-digit telephone dialing. Although the investigators observed an association between certain types of wiring configurations, no clear association was seen between leukemia risk and the magnetic or electric fields that were measured (8).

The case-control study design is highly efficient. It requires a relatively short time between implementation and identification of the quantitative outcome. Furthermore, because most diseases are relatively rare events, the design is efficient in requiring that relatively fewer persons be contacted to determine whether a relationship exists. Although the design is efficient, it remains susceptible to a variety of biases or systematic errors.

The systematic errors most frequently associated with case-control studies include recall bias and selection bias. Recall bias implies that persons with the disease are more likely to have differential recall of a past exposure after receiving their diagnosis. The potential for selection bias occurs when the control group is not constructed randomly and does not have the same likelihood of disease (other than disease resulting from the exposure of interest) as that of the population that gave rise to the cases.

An example of selection bias is the "healthy worker effect", observed in occupational epidemiology. Essentially, the healthy worker effect refers to the observation that employed persons tend to be healthier than the general population, which includes those too ill to work (*see* Figure 3). As people become ill, they tend to remove themselves from certain occupations. In environmental epidemiology, one sees a similar phenomenon in which people who fear an environmental contamination relocate to geographical regions where they perceive less contamination. Likewise, people with resources have greater flexibility in minimizing their exposure to localized environmental contamination. Thus, individuals who become study subjects in a localized contamination may be different from the general population.

Case-control studies also offer the opportunity to "nest" themselves within other study designs. This approach is used when a case-control study is a component of a cohort study or in conjunction with a surveillance program. The leukemia and magnetic fields example represents the use of a nested case-control study in which cases were identified by means of an ongoing tumor surveillance program.

Intervention Studies. Intervention studies are found infrequently among epidemiological studies of the environment. There may be surveillance studies that examine air, water, or soil quality before and after an intervention. However, rarely is the health of humans examined before and after implementing an environmental intervention.

$$\text{Cohort Study:} \quad P \frac{\frac{d \text{ (exposed)}}{\bar{d} \text{ (exposed)}}}{\frac{d \text{ (unexposed)}}{\bar{d} \text{ (unexposed)}}} = \frac{\text{incidence (exposed)}}{\text{incidence (unexposed)}} = \text{relative risk}$$

$$\text{Case-Control Study:} \quad P \frac{\frac{e \text{ (diseased)}}{\bar{e} \text{ (diseased)}}}{\frac{e \text{ (non-diseased)}}{\bar{e} \text{ (non-diseased)}}} = \text{estimate of RR} = \text{Odds Ratio}$$

Figure 2. *The "healthy worker effect" can generate systematic error in associations between disease and exposure in cross-sectional studies.*

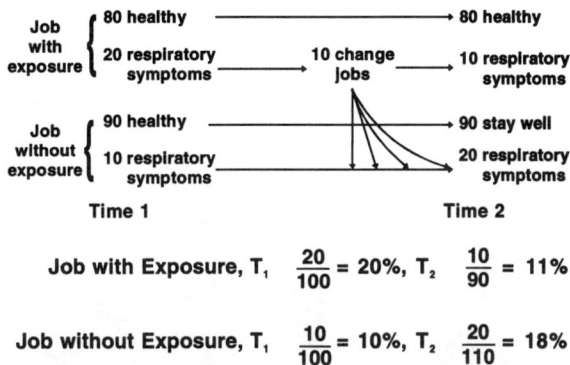

Figure 3. *A comparison of the estimators of risk from the cohort and case-control study designs. (RR = relative risk.)*

Summary

Study designs are tactics or strategies for eliciting the nature of a relationship between an exposure and an outcome within a given population. The selection of a particular study design is based on the investigator's degree of confidence in the time order and directionality of the relationship. Pragmatically, frequency of disease occurrence, cost of research, ethics of randomizing populations to exposure, and availability of natural experiments are factors that the investigator should consider in selecting a study design.

References

1. Hennekens, C. H.; Buring, J. E. In *Epidemiology in Medicine*; Mayren, S. L., Ed.; Little, Brown: Boston, MA, 1987.
2. Ahlbom, A.; Norell, S. *Introduction to Modern Epidemiology*; Epidemiology Resources: Chestnut Hill, PA, 1990.
3. Hulley, S. B.; Cummings, S. R. *Designing Clinical Research: An Epidemiologic Approach*; Williams & Wilkins: Baltimore, MD, 1988.
4. Checkoway, H.; Pearce, N. E.; Crawford-Brown, D. J. *Research Methods in Occupational Epidemiology*; Oxford University: New York, 1989.
5. Monson, R. R. *Occupational Epidemiology*, 2nd ed.; CRC: Boca Raton, FL, 1990.
6. Centers for Disease Control. "Guidelines for Investigating Clusters of Health Events". *Morbidity, Mortality Weekly Rep.* **1990**, *39(RR-11)*, 1–17.
7. Sowers, M. F. R.; Clark, M. C.; Jannausch, M. L.; Wallace, R. B. *Am. J. Epidemiol.* **1991**, *133*, 649–660.
8. London, S. L.; Thomas, D. C.; Bowman, J,; Sobel, E.; Cheng, T. L.; Peters, J. M. *Am. J. Epidemiol.* **1991**, *134*, 923–937.

RECEIVED for review September 3, 1992. ACCEPTED revised manuscript December 23, 1992.

4

Interpreting Epidemiological Studies

George Maldonado

Division of Environmental and Occupational Health, School of Public Health, University of Minnesota, Minneapolis, MN 55455

> *A primary objective of an epidemiological study is to obtain a valid (unbiased) estimate of the effect of an exposure on disease occurrence. Most epidemiological studies are observational (nonexperimental) and consequently are subject to more potential biases than experiments. This chapter (1) discusses and gives examples of the major potential biases that may occur in epidemiological studies, (2) discusses strategies for controlling each bias, and (3) discusses how to interpret epidemiological results when strategies for controlling bias fail to achieve full control.*

INVESTIGATORS IN MOST SCIENTIFIC disciplines employ an experimental design in which subjects are randomly allocated to different exposure groups, many of the study conditions are precisely controlled, and relevant factors are precisely measured.

Epidemiologists rarely have this luxury. Most epidemiological studies use an observational (nonexperimental) design. Instead of randomly allocating subjects to different levels of an exposure whose effects the investigator wants to study, epidemiologists observe groups of people and attempt to ascertain accurately the level of exposure that each person has experienced. Instead of precisely controlling experimental conditions, epidemiologists observe extraneous factors that might influence the study outcome, and they attempt to control them with observational designs and statistical analysis techniques. Instead of precisely measuring relevant factors, epidemiologists frequently "make do" with imprecise measurements and the imperfect memory of subjects.

Because of the observational nature of most epidemiological studies, the estimates one computes from epidemiological data may suffer from bias (systematic deviation of results from truth). Therefore, to correctly interpret epidemiological studies, one must

1. recognize and understand the important potential sources of bias
2. evaluate the magnitude and direction of potential biases

In other words, interpreting the results of an epidemiological study is like shaping a rough diamond into a finished product. The rough diamond is like the potentially biased estimates of effect one obtains from an observational epidemiological study. Shaping the diamond is like making inferences about the magnitude and direction of potential biases. The finished stone is like the combination of estimates of effect and inferences about biases.

Making inferences about the magnitude and direction of possible biases is not a simple task. It takes much knowledge and experience to "shape the diamond". This chapter outlines a strategy for making inferences about potential biases and highlights the important issues to be considered. Rothman (1), Kelsey et al. (2), Checkoway et al. (3), and Greenland (4) give more indepth discussions of these issues.

This chapter assumes a familiarity with basic epidemiological terminology such as relative risk, odds ratio, and risk factor. It also assumes a basic understanding of cohort and case-control epidemiological study designs. The two preceding chapters in this volume and Ahlbom and Norell (5) or Walker (6) give an introduction to these issues.

A Basic Strategy for Interpreting Epidemiological Studies

The three most important types of bias that can occur in epidemiological studies are the following:

1. information bias
2. confounding
3. selection bias

A basic strategy for interpreting epidemiological results is to evaluate the magnitude and direction of each of these three biases. This chapter (1) defines and gives examples of these three biases, (2) discusses strategies for controlling each bias, and (3) discusses how to interpret epidemiological results when bias may be present.

Information Bias

Bias due to errors in measuring (or classifying) the study variables is called information bias.

Information bias is perhaps the most important bias in epidemiological studies, for several reasons:

1. There are sources of measurement error in nearly all epidemiological studies. For example, measurement error may be caused by imperfect recall of subjects, by improper calibration of measurement equipment, and by use of a proxy variable as a substitute for the actual variable of interest.
2. Nearly all sources of measurement error cause information bias.
3. The magnitude of information bias can be large.
4. Information bias can usually be handled only by minimizing measurement error or by making inferences about the magnitude and direction of bias. Unlike some of the other biases, it usually cannot be eliminated by data analysis techniques.

Hypothetical Example. Consider a case-control study of an occupational chemical exposure and lung cancer. Assume that if all subjects were correctly classified into "exposure" and "disease" categories, an investigator would observe the following data:

Perfectly Classified Data (Truth)

Exposure Category	Case	Control
Exposed	200	200
Unexposed	50	200

The *correct* (true) odds ratio (OR) is

$$\text{OR} = \frac{200 \times 200}{50 \times 200} = 4.0.$$

In this hypothetical example, exposure to the chemical multiplies the risk of disease by a factor of 4.

Assume also the following:

1. Data on actual chemical exposure are not available, so subjects are classified into exposure categories on the basis of their job titles.
2. As a result of this imperfect method of determining exposure, 50% of exposed subjects are misclassified into the "unexposed" group, and all unexposed are correctly classified.

Because of the foregoing assumptions, the investigator would observe the following data:

Misclassified Study Data

Exposure Category	Case	Control
Exposed	100	100
Unexposed	150	300

Using the misclassified study data, the estimated OR is

$$OR = \frac{100 \times 300}{150 \times 100} = 2.0$$

which is not equal to the true OR. Information bias makes it appear that exposure to the chemical multiplies the risk of disease by a factor of 2 (instead of the true factor of 4).

Evaluating the Magnitude and Direction of Information Bias. To evaluate the magnitude and direction of information bias, the sources of measurement (or classification) error must be identified. The magnitude of the bias depends on the magnitude of the error in measuring (or classifying) variables. The greater the rate of error, the greater the bias. The direction of the bias depends on how the measurement (or classification) error is distributed in the data. (A discussion of these issues was presented by Copeland et al. (7), Greenland (8), Greenland and Robins (9), Poole (10), Dosemeci et al. (11), and in the letters in the August 15, 1991, issue of the *American Journal of Epidemiology*.)

Confounding

The most meaningful effect estimates one obtains from epidemiological data have two characteristics: (1) they directly compare the occurrence of disease in two groups that have different levels of an exposure; (2) they isolate the effect of the exposure from other factors that might also influence the disease of interest—factors to be controlled, not studied. For example, an investigator might want to estimate how the rate of disease increases as some occupational exposure increases, adjusted for the effects of other exposures or personal characteristics that also affect the disease of interest. These comparison estimates usually take the form of ratios or differences of disease rates or risks (e.g., "relative risk" and "attributable risk" measures).

When one computes these comparison estimates, a fundamental assumption is made: One assumes that disease occurrence among the unexposed accurately predicts what would have happened in the exposed group if they had not been exposed. The central idea is this: One assumes that in the absence of exposure, disease occurrence would be the same in both groups (12, 13).

If this assumption is incorrect, the observed comparison between exposure groups is confounded (i.e., confounding exists). That is, the estimate of effect reflects not only the effect of exposure but also the effects of other factors that influence disease occurrence (i.e., factors to be controlled, not studied). The effect estimate will be a biased measure of the impact of the exposure alone on disease occurrence.

How can one tell if confounding is present? Answer this question: Would disease occurrence be the same in the two groups if the exposure was absent? If the answer is no, then confounding is present. To find the answer, we search for differences between exposure groups in the distributions of extraneous risk factors for the disease (i.e., risk factors to be controlled, not studied). We search for confounders.

Hypothetical Example. Consider a fixed-cohort study of the relationship of nitroglycerin (an occupational chemical exposure) to death from myocardial infarction (MI).

Let us assume the following:

1. Regular exercise protects against MI death.
2. Of the exposed subjects, 95% exercise regularly, whereas only 50% of the unexposed subjects exercise regularly.

In this hypothetical example, exercise confounds a simple comparison of nitroglycerin-exposed versus -unexposed groups. If nitroglycerin exposure had no effect on MI death, we should *not* expect the occurrence of MI deaths to be the same in the exposed and unexposed groups. In fact, we should expect that MI deaths will occur *less* frequently among exposed subjects [because of assumptions (1) and (2)].

The following is a hypothetical data set with exercisers and nonexercisers combined:

Exercisers and Nonexercisers Combined

Exposure Category	Deaths	Nondeaths	Total
Exposed	1200	208,800	210,000
Unexposed	1250	198,750	200,000

The biased (confounded) relative risk (RR) is

$$RR = \frac{1200/210,000}{1250/200,000} = 0.9$$

It appears that nitroglycerin exposure multiplies the risk of MI death by a factor of 0.9. (i.e., nitroglycerin exposure is slightly protective of MI death). This estimate is biased because of the confounding effect of exercise.

If we examine exercisers and nonexercisers separately (i.e., if we control for confounding by exercise), we can compute unconfounded estimates of effect:

Exercisers

Exposure Category	Deaths	Nondeaths	Total
Exposed	1000	199,000	200,000
Unexposed	250	99,750	100,000

$$RR = \frac{1000/200,000}{250/100,000} = 2.0$$

Nonexercisers

Exposure Category	Deaths	Nondeaths	Total
Exposed	200	9800	10,000
Unexposed	1000	99,000	100,000

$$RR = \frac{200/10,000}{1000/100,000} = 2.0$$

The correct RR is 2.0 for both exercisers and nonexercisers. Exposure to nitroglycerin doubles the risk of MI death. The exposure is not protective of MI death as the confounded estimate would have led us to believe.

Control of Confounding. For confounding factors that have been measured, an investigator can use study design and data analysis techniques to eliminate confounding. Study design techniques include the following:

1. randomly allocating subjects to exposure groups (experiment) (This strategy also helps control unmeasured confounders.)
2. restricting the eligibility of subjects according to values of the potential confounders
3. matching in a cohort study (i.e., selecting subjects so that exposed and unexposed groups have similar distributions of potential confounding factors)

Data analysis techniques include the following:

1. stratified analysis methods
2. multivariate modeling techniques

Judging the Magnitude and Direction of Confounding. To evaluate confounding due to unmeasured confounders, an investigator must first identify potential confounding factors that were not measured. A discussion of this issue was given by Miettinen and Cook (12), Greenland and Robins (13), Rothman (1), and Kass and Greenland (14).

The magnitude of the bias depends on how strongly the confounder affects disease occurrence and on how strongly the confounder and the exposure are associated. The stronger the confounder–disease relationship and the stronger the exposure–confounder association, the greater the bias will be. The direction of the bias depends on the directions of these associations. A more detailed discussion was given by Flanders and Khoury (15).

Selection Bias

In a case-control study, bias due to the way in which cases or controls are selected for study is called selection bias.

A case-control study can yield an unbiased estimate of the effect of exposure on disease occurrence if the following two conditions are met (1, 4):

1. The sample of cases gives an unbiased estimate of the exposure distribution among cases in the source population over the study period.
2. The sample of controls gives an unbiased estimate of the exposure distribution in the population at risk over the study period.

If these two conditions are not met because of the way cases or controls are selected for study, selection bias may exist, and estimates of effect may be incorrect.

Unfortunately, biased samples of cases and controls can occur for many reasons, which is why selection bias should be considered when interpreting epidemiological studies. For example, if exposed cases refuse to participate more often than unexposed cases, the studied cases will not provide an unbiased estimate of the exposure distribution among cases in the population during the study period. For another example, if all exposed cases are detected, but some unexposed cases are undetected, the cases studied will not provide an unbiased estimate of the exposure distribution among cases in the population during the study period. For a final example, if exposed controls refuse to participate more often than unexposed controls, the controls studied will not provide an unbiased estimate of the exposure distribution among the population at risk during the study period.

Hypothetical Example. Consider the following hypothetical population data for the relationship between a chemical exposure and lung cancer:

Source Population (Truth)

Exposure Category	Case	Noncase
Exposed	400	9600
Unexposed	600	14,400

In this population, OR is

$$OR = \frac{400 \times 14{,}400}{600 \times 9600} = 1.0$$

Exposure to the chemical is not a risk factor for lung cancer in the source population.

Consider a case-control study of this population. Assume that, unknown to the investigator, subjects were sampled into the study with the following selection probabilities:

Probabilities of Being Selected into Study

Exposure Category	Case	Noncase
Exposed	0.4875	0.0146
Unexposed	0.2583	0.0146

Here are the data one would expect with the foregoing selection probabilities:

Study Data

Exposure Category	Case	Control
Exposed	195	140
Unexposed	155	210

The exposed and unexposed controls have equal selection probabilities, and so controls will give an unbiased estimate of the exposure distribution among the population at risk. In the population, 9600/24,000 = 0.40 of the noncases were exposed to the chemical. From the controls in the case-control data, we would correctly estimate that 140/350 = 0.40 of the population at risk was exposed to the chemical.

The exposed and unexposed cases have *unequal* selection probabilities, perhaps because unexposed cases refused to participate in the study more often than exposed cases. Because of the unequal selection probabilities, the cases in the study *will not* give an unbiased estimate of the exposure distribution among cases in the population. In the population, 400/1000 = 0.40 of the cases were exposed to the chemical. From the cases in the case-control study, we would incorrectly estimate that 195/300 = 0.56 of the cases in the population were exposed to the chemical.

The biased OR,

$$OR = \frac{195 \times 210}{155 \times 140} = 1.9,$$

is not equal to the true OR (1.0) because of selection bias. Selection bias would lead us to conclude incorrectly that chemical exposure increases the risk of lung cancer.

Controlling Selection Bias. If we can identify (and measure) a factor that affects the chance of selection into the study, we can adjust for this factor in the analysis and can remove selection bias.

Evaluating the Magnitude and Direction of Selection Bias. Making inferences about the magnitude and direction of selection bias is difficult. First an investigator must identify reasons why the sample of cases or the sample of controls is biased—in practice, this is exceedingly difficult to do. One can then perform an evaluation using sensitivity analysis (i.e., a "What if?" analysis based on how one thinks selection might be biased) [e.g., Maclure and Hankinson (*16*)].

Summary

Interpreting the results of an epidemiological study is like shaping a rough diamond into a finished gem. The rough diamond is like the potentially biased estimates of effect one obtains from an observational epidemiological study. Because of the observational nature of most epidemiological studies, results are almost always biased to some degree. Shaping the diamond is like making inferences about the magnitude and direction of potential biases. The most important biases in epidemiological studies are information bias, confounding, and selection bias. The finished stone is like the combination of estimates of effect and inferences about biases.

Acknowledgments

The author thanks Sander Greenland, Jack Mandel, and the students of PubH 8194, Spring Quarter, 1992, University of Minnesota School of Public Health, for their comments.

References

1. Rothman, K. J. *Modern Epidemiology*; Little, Brown: Boston, MA, 1986.
2. Kelsey, J. L.; Thompson, W.D.; Evans, A. S. *Methods in Observational Epidemiology*; Oxford University: New York, 1986.
3. Checkoway, H.; Pearce, N. E.; Crawford-Brown, D. J. *Research Methods in Occupational Epidemiology*; Oxford University: New York, 1989.
4. Greenland, S. In *Oxford Textbook of Public Health*, 2nd ed.; Holland, W. W.; Detels, R., Eds.; Oxford University: New York, 1991; Vol. 2, Methods of Public Health.
5. Ahlbom, A.; Norell, S. *Introduction to Modern Epidemiology*; Epidemiology Resources: Chestnut Hill, PA, 1990.
6. Walker, A. M. *Observation and Inference*; Epidemiology Resources: Chestnut Hill, PA, 1991.
7. Copeland, K. T.; Checkoway, H.; McMichael, A. J.; Holbrook, R. H. *Am. J. Epidemiol.* **1977**, *105*, 488–495.
8. Greenland, S. *Am. J. Epidemiol.* **1980**, *112*, 564–569.
9. Greenland, S.; Robins, J. M. *Am. J. Epidemiol.* **1985**, *122*, 495–506.
10. Poole, C. *Am. J. Epidemiol.* **1985**, *122*, 508.
11. Dosemeci, M.; Wacholder, S.; Lubin, J. H. *Am. J. Epidemiol.* **1990**, *132*, 746–748.
12. Miettinen, O. S.; Cook, F. *Am. J. Epidemiol.* **1981**, *114*, 593–603.
13. Greenland, S.; Robins, J. M. *Int. J. Epidemiol.* **1986**, *15*, 413–419.
14. Kass, P. H.; Greenland, S. *J. Am. Vet. Med. Assoc.* **1991**, *199*, 1569–1573.
15. Flanders, W. D.; Khoury, M. J. *Epidemiology* **1990**, *1*, 239–246.
16. Maclure, M.; Hankinson, S. *Epidemiology* **1990**, *1*, 441–447.

RECEIVED for review September 3, 1992. ACCEPTED revised manuscript December 22, 1992.

5

Environmental Epidemiology for Chemists

Curtis D. Klaassen[1] and William C. Kershaw*[,2]

[1]University of Kansas, School of Medicine, Kansas City, KS 66103
[2]Health Care Technology Division, Preclinical Safety Assessment, Procter and Gamble Company, Miami Valley Laboratories, Cincinnati, OH 45239-8707

> *The principles of toxicology are utilized to (1) characterize the adverse effects of chemicals on living organisms, (2) define the conditions of exposure required to produce these pernicious effects, and (3) understand the mechanisms by which chemicals produce toxicity. Toxicologists conduct safety studies to determine the intrinsic toxic properties of chemicals and conduct pharmacokinetic studies to assess the biological disposition of chemicals. The most important tenet of toxicity testing is the dose–response relationship, because conclusions regarding causality between chemical exposure and the ensuing adverse reaction are based on it. Toxicologists obtain essential information from safety and pharmacokinetic studies to assess the risks associated with chemical exposure and, in this manner, protect human health from hazardous materials found in the environment, the workplace, and the home.*

Introduction to Toxicology

Toxicology is the study of the adverse effects of chemicals on living organisms. The toxicologist is an individual trained to examine the nature of these adverse effects and to determine the probability of their occurrence. Since the 1970s, several academic institutions have offered specialized training in this field, awarding both undergraduate and graduate degrees in toxicology. In contrast to this recent trend of specialized didactic training, many practicing toxicologists with graduate degrees in physical and life sciences have acquired a working understanding of toxicology by applying these sciences to toxicology-related issues in industry, academia, and government.

*Corresponding author

Different Areas of Toxicology. Toxicologists apply their skills to three major fields of inquiry: descriptive, mechanistic, and regulatory toxicology. Descriptive toxicology is a process whereby the deleterious effects of chemicals are characterized by using laboratory animals and various in vitro techniques. The purpose of these studies is to define the hazards that occur during circumscribed exposure conditions. In the pharmaceutical industry, one of the goals of the descriptive toxicologist is to determine the no observed adverse effect level (NOAEL) of a drug candidate in order to set safe dose levels for first-time exposure in humans. In the chemical industry, the toxicologist is concerned not only about human health hazards posed by company products but also about the potential deleterious effects these chemicals may have on the environment and its inhabitants. In mechanistic toxicology, the mechanisms by which chemicals exert their toxic effects are determined in the laboratory. Learnings from this area of study frequently lead to the development of more efficient safety tests for the prediction of human health hazards, the establishment of structure–activity relationships for the design of molecules with improved safety profiles, the identification of antidotes to treat poisonings, and a clearer understanding of human physiology and disease processes. Regulatory toxicology is an activity conducted by governmental agencies whereby the hazards posed by industrial chemicals or pharmaceuticals are assessed, and judgments are made about whether the product can be utilized in the manner indicated by the sponsoring company. Regulatory toxicologists base their hazard assessments on data provided by the descriptive toxicologist employed by the company that is petitioning the governmental agency for marketing approval.

Toxicity versus Hazard. The principle activity of descriptive and regulatory toxicologists is the assessment of the hazards or risks associated with the use of chemicals and not the intrinsic toxic activity of chemicals per se. Hazard is the situation in which an undesirable response results from a specified level of exposure. For example, tetrodotoxin is an extremely toxic substance and is present in the liver of puffer fish, whose flesh is considered a delicacy in Japan. Although a single dose of tetrodotoxin as low as 0.1 mg/kg is lethal to laboratory animals, this substance is not a major human health hazard, because the fish are prepared by experienced food handlers who are careful not to contaminate the flesh with liver. By comparison, ethanol is not as intrinsically toxic as tetrodotoxin. A lethal dose of ethanol is approximately 100,000 times greater than that of tetrodotoxin. Despite the fact that ethanol is tremendously less toxic than tetrodotoxin, one could convincingly argue that ethanol is more hazardous, in view of the number of alcohol-related deaths on our nation's highways each year. Both intrinsic toxicity and exposure conditions (i.e., route, duration, frequency and probability of exposure, and dose) determine whether a substance will be hazardous.

Local versus Systemic Effects. Toxic reactions may result from a direct local action or from the absorption of the toxicant into systemic circulation and its subsequent distribution into various tissues. Local effects refer to those that occur at the site of first contact between the toxicant and the organism. Most substances, except for highly reactive ones, such as very caustic or irritating substances, require systemic absorption and tissue distribution to produce their toxic effects. These categories are not mutually exclusive, because some substances demonstrate both local and systemic effects. For instance, tetraethyl lead damages skin on contact and after systemic absorption causes central nervous system intoxication manifested by restlessness, headaches, loss of memory, and convulsions.

Spectrum of Undesired Effects. The spectrum of undesired effects produced by systemically active toxicants is wide. Toxicities run the gamut from relatively benign or bothersome to incapacitating and life-threatening. Drugs, for instance, often produce a wide array of subtle changes in bodily functions, although the prescribing physician often seeks to produce a single effect. Results not directed toward the intended therapeutic objective are considered undesirable effects, or side effects, of the drug for a particular disease condition. These untoward effects are usually not of major concern; in fact, an undesirable effect for one therapeutic indication may be desirable for another clinical situation. For example, atropine causes dryness of the mouth, which is considered a side effect when the drug is prescribed for the treatment of peptic ulcers but is the desired response when it is given as a preanesthetic medication.

More serious toxic reactions that result from systemic exposure can be characterized by the onset of toxicity following exposure, by whether the deleterious changes are permanent or temporary, by a genetically regulated abnormal reactivity of the exposed organism to the toxic agent, and by whether the adverse reaction results from an immune response.

Immediate versus Delayed Toxicity. Immediate toxic reactions are those that develop within hours or days following chemical exposure, whereas delayed effects become apparent weeks or even years after the exposure incident. Carbon tetrachloride-induced hepatotoxicity is an example of immediate toxicity, because marked changes in the biochemistry and structure of rat liver are observed several hours after injection of toxic doses. In contrast, tumors often appear 20 years or more after exposure to a carcinogen. For instance, there are numerous cases in which young women have developed vaginal and uterine tumors as the result of in utero exposure to diethylstilbestrol when this drug was prescribed to their mothers to prevent miscarriages.

Reversible versus Irreversible Toxic Effects. Damage to tissues caused by chemical exposure can be either reversible or irreversible. The severity of chemically mediated tissue damage and the type of tissue in which injury occurs are important factors that dictate whether toxic effects will be temporary or become permanent. For example, liver has a remarkable capacity to regenerate after chemical insult. Consequently, the cytotoxic effects of many hepatotoxicants are completely reversible. Chemically mediated damage to the central nervous system, however, is largely irreversible because fully differentiated neural tissue has relatively little ability to self-repair.

Idiosyncratic Reactions. "Idiosyncrasy" is defined as a genetically controlled, unexpected reaction to a chemical agent. The response that follows exposure is qualitatively similar to that observed in most individuals, but the intensity of the response is considerably greater or less than expected. Idiosyncrasy can be illustrated by the unanticipated protracted duration of apnea (i.e., cessation of breathing) that is observed in 1 in 2500 patients who receive a standard dose of the skeletal muscle relaxant, succinylcholine. In most people, succinylcholine produces transient apnea because of its rapid degradation and inactivation by plasma cholinesterase. Patients who exhibit this idiosyncratic reaction have an atypical form of plasma cholinesterase that has a reduced ability to metabolize the drug and thereby to neutralize its activity. The familial occurrence of this phenomenon suggests that hypersusceptibility to succinylcholine is under genetic control.

Allergic or Sensitization Reactions. Chemical allergic reactions are immunologically mediated and require a previous period of exposure (i.e., sensitization) to the chemical or to a structurally similar one. One of the hallmarks of chemical allergies is a heightened sensitivity to chemical intoxication in preexposed individuals. A population is composed of "naive" people (i.e., nonexposed individuals) as well as people who have had various degrees of prior contact with the chemical. Consequently, there is marked person-to-person variation with respect to the susceptibility to allergic chemicals. Therefore, it is not surprising that population-based, dose–response relationships are often not readily apparent. The apparent lack of a dose–response relationship suggests that allergic reactions are not true toxic responses. However, a dose–response relationship can be demonstrated in a single allergic individual, and the allergic responses are often severe and occasionally life-threatening. Therefore, allergic reactions should be regarded as a genuine toxic response.

Systemic Effects. In descriptive toxicity studies, chemicals are administered via routes and dose forms that maximize systemic absorption. The goal of these studies is to characterize fully a compound's ability to produce biochemical, structural, and functional changes in various target organs, such as liver, kidney, lung, bone marrow, and brain. Compounds are also tested as to their ability to (1) interact with DNA and alter the process of heredity (i.e., mutagenesis), (2) produce tumors (i.e., carcinogenesis,), (3) reduce the fertility of male and female animals (i.e., reproductive toxicity), and (4) cause malformation and dysfunction of offspring (i.e., developmental toxicity).

The Dose–Response Relationship. To establish that the administration of a chemical is causally related to signs of intoxication, the toxicologist must demonstrate a dose–response relationship. This relationship is based on the reasonable presumption that the toxic response is a result of the administered chemical and the assumption that the intensity of this response is related proportionally to the dose. The assumption is based on three others: First, there is a molecular site within the organism with which the chemical interacts to produce the response; second, the intensity of the response is related to the concentration of the chemical at that site; and third, the concentration at the site is related to the dose. Dose–response curves can be either "graded" or "quantal". The dose describes the increasing intensity of a response of an individual (or a population) over a range of increasing dosages, whereas the response illustrates the distribution of minimal doses that produce a given "all-or-none" response (e.g., presence of a tumor) in a population of biological subjects.

Organizations That Conduct Toxicity Studies. The pharmaceutical industry is responsible for conducting descriptive animal toxicity studies that determine which organs are the potential targets of toxicity during drug therapy as well as the drug's mutagenic, carcinogenic, and teratogenic properties. The Food and Drug Administration (FDA) is responsible for reviewing animal toxicology reports submitted by the sponsoring drug company and for judging whether the potential benefits of the drug outweigh its accompanying hazards. If a favorable risk-to-benefit paradigm is concluded, the drug candidate is approved for initial clinical trials that establish safety and tolerance in humans. The Environmental Protection Agency (EPA) requires the chemical industry to conduct extensive descriptive toxicity testing of pesticides and industrial chemicals. The resulting data are used by the EPA to assess the possible negative impact a chemical may have on the environment and to determine whether remedial activities against the manufacturer should be initiated.

Pharmacokinetics and Toxicity Studies

Two fundamental types of studies, broadly referred to as pharmacokinetic and toxicity studies, are conducted to understand the interaction between chemicals and living organisms. Pharmacokinetic studies determine the body's effects on the chemical, whereas toxicity studies determine the effects of the chemical on the body.

Pharmacokinetic Studies. Strictly speaking, manifestations of systemic intoxication do not depend directly on the administered dose of a toxicant but rather on the concentration of the chemical (and its metabolites) in target organs. The concentration of the active chemical species in target organs is regulated by four biological processes: absorption, distribution, metabolism, and elimination.

Absorption. The process by which chemicals cross biological membranes and enter blood is called absorption. The gastrointestinal tract, lungs, and skin provide pathways for chemicals to enter systemic circulation. Both the rate and the extent with which toxicants enter blood markedly influence the severity and type of resulting lesion. Injection of chemicals into veins or the peritoneal cavity (for rodents only) are additional routes of exposure that are employed in the laboratory to evaluate their pharmacokinetic and toxicological properties.

The gastrointestinal tract is an important site of absorption through which foreign chemicals gain access to systemic circulation. For example, many environmental pollutants enter the food chain and are ingested along with food products. This route of exposure is also encountered frequently in suicide attempts involving overdoses of drugs and is the most common route in which children are poisoned by household chemicals.

Absorption after oral ingestion is influenced by physicochemical properties of the chemical. Factors that favor absorption include high lipid-to-water partition coefficient, low degree of ionization at various pH levels along the gastrointestinal tract, low molecular weight and rapid dissolution rate for solid materials. Properties of the exposed individual, such as gastric emptying time, intestinal motility, splanchnic blood flow, bile secretion, and binding to gastrointestinal mucus, as well as interactions between chemicals and foods also have a significant influence on the absorption of foreign substances that are ingested unintentionally.

Distribution. Following absorption, most xenobiotics are carried by blood in a nonuniform manner throughout the body. Whether a chemical remains in systemic circulation, diffuses into extracellular fluid, or enters and is stored in various tissue compartments, such as adipose tissue and bone, depends on its physicochemical properties (i.e., lipophilicity, degree of ionization at plasma pH, molecular size, and permeability through plasma and tissue membranes) as well as biological considerations, such as binding to plasma proteins and cellular constituents. Of particular interest to the toxicologist is the propensity of some toxicants to accumulate in specific tissues in the body. These storage depots often are the target organs of toxicity; for example, paraquat is a pulmonary toxicant that accumulates in lung. Conversely, some toxicants accumulate in storage depots that are not target tissues; sequestration in these tissues can be considered a protective mechanism. For example, environmental contaminants such as lead and 1,1,1-trichloro-2,2-bis(*p*-chlorophenyl)ethane (DDT) distribute and concentrate in bone and body fat, respectively. These compounds have no toxicological impact on their respective storage sites and once sequestered are not available to affect their target organ, the central nervous system.

Metabolism. Most chemicals undergo enzymatically catalyzed biotransformations that produce metabolites that are more polar and water-soluble than the parent compound. Increasing the water-soluble properties of a compound enhances its elimination from the body and thereby lowers the body burden of the material. Biotransformation reactions can convert a toxic chemical to an innocuous metabolite or to a metabolite that has similar or greater toxicologic activity than the parent compound. The liver is the principal site of xenobiotic metabolism, although biotransformation also occurs in kidney, intestinal epithelium, lung, blood, and in the lumen of the gut by the action of intestinal flora.

Biotransformation reactions are classified as Phase I and Phase II reactions. Phase I reactions usually convert the parent compound to more polar metabolites and involve three types of chemical reactions: (1) oxidation (e.g., hydroxylation, side-chain aliphatic oxidation, deamination, sulfoxide formation, dealkylation of O, N, and S substitutions, and the dehydrogenation of alcohols and aldehydes), (2) reduction (e.g., azo, nitro, ketone, and aldehyde), and (3) the hydrolysis of esters and amides.

Phase II reactions are conjugation or synthetic reactions. One of the most common conjugation reactions is catalyzed by uridine diphosphoglucuronosyltransferase and results in the formation of glucuronic acid derivatives. Various chemical moieties are susceptible to glucuronidation, such as aliphatic and aromatic amines as well as hydroxyl, carboxyl, and sulfhydryl groups. Other conjugation reactions include the acetylation of primary amines; conjugation with sulphate, glycine, and glutathione; and methylation of O, N, and S atoms in the parent molecule.

Phase I and Phase II reactions usually result in the formation of metabolites that are more readily excreted in the urine. Some Phase II reactions, such as conjugation with glucuronic acid or glutathione, frequently shift the route of excretion from urine to feces, as described further on in greater detail.

Elimination. Parent compound and its metabolites are eliminated from the body via urine and feces and, to a lesser extent, by way of respiration and secretory substances (i.e., milk, sweat, saliva, and tears). Excretory organs typically eliminate polar or water-soluble compounds more proficiently than they do lipophilic compounds. Thus, xenobiotic metabolism frequently enhances the elimination of chemicals and so can be considered a detoxification process.

Kidney, the primary organ for chemical elimination, receives 25% of the cardiac output, which allows large volumes of plasma water to be filtered. Several factors favor glomerular filtration of chemicals, such as low molecular weight; high concentration gradient between plasma water and filtrate; and limited plasma protein binding, because only unbound chemical can be filtered. Once chemicals are present in tubular filtrate, the extent of ionization influences renal excretion, because polar compounds and ions are not reabsorbed in proximal or distal tubules. Nonionized compounds having high lipid-to-water partition coefficients are passively reabsorbed through tubular cell membranes. Weak acids are nonionized in acidic urine, which promotes recycling and prolongs elimination. Conversely, alkalinization of urine accelerates the excretion of weak acids. Biotransformation reactions frequently result in the formation of stronger acids that readily ionize in acidic urine. This effect reduces tubular reabsorption, facilitates urinary excretion, and often decreases a toxicant's duration of activity.

Energy-dependent, saturable transport of organic acids and bases from afferent blood into tubular urine occurs by way of proximal tubular cells and is more effective than filtration as a means to eliminate compounds. The fact that only 20% of afferent blood that flows to the kidney is filtered, whereas the remaining 80% is delivered to tubular cells, partially accounts for the effectiveness of secretion. Reversible binding of chemicals to plasma proteins does not significantly affect proximal tubule secretion, because free and bound chemicals are available for transport. This occurs because bound chemical dissociates rapidly to maintain equilibrium with plasma water as free chemical is removed from tubular cells.

Chemicals can also be excreted by hepatocytes across canalicular membranes into the biliary tract and pass into intestine. Low-molecular-weight molecules are poorly excreted into bile, whereas compounds with molecular weights of 350 daltons or more are excreted into bile in appreciable quantities. The synthetic nature of Phase II reactions results in metabolites with higher molecular weights than those of the parent chemical. Thus, the products of conjugation reactions that use glutathione and glucuronic acid as ligands are typically excreted into bile. Once in the intestine, the chemical and its metabolite can be eliminated in the feces. Alternatively, if the properties of the compound favor intestinal absorption (i.e., low degree of ionization, high lipid-to-water partition coefficient of the nonionized form, and small molecular radius of the water-soluble substance); a cycle referred to as "enterohepatic circulation" may result, in which biliary secretion and intestinal reabsorption continue until biotransformation and urinary excretion eliminates the compound from the body. Glucuronide conjugates are susceptible to hydrolytic reactions conducted by gut flora. These reactions form products that have physicochemical properties favorable to intestinal reabsorption. Sometimes a chemical's long persistence in the body is largely the result of enterohepatic circulation, as is the case with cardiac glycosides. In contrast, biliary excretion serves as an efficient route of elimination for certain organic acids and bases (e.g., tetracyclines and streptomycin) that cannot be reabsorbed in the intestine, owing to ionization at intestinal pH.

Toxicology Studies. Hazard Evaluation Goals. The purpose of descriptive toxicity studies, regardless of the intended use of the test material, is to determine the adverse effects produced by chemicals and the dose needed to produce these effects. Ultimately, this vital information is used by the toxicologist to make hazard assessments to protect humans against the detrimental effects of chemicals.

Principles of Toxicity Testing. Descriptive toxicity testing is founded on two main principles that support its use in chemical hazard assessment. The first principle is that the toxic effects produced by a compound in laboratory animals, when properly qualified, are applicable to humans. This tenet is not unique to toxicology and applies to all other life sciences. Furthermore, it has been verified numerous times for a wide variety of chemicals. For example, nearly all human carcinogens are also carcinogenic in at least one species used in toxicity studies.

The second principle is that exposure of experimental animals to high doses of toxic agents is necessary and a valid method of discovering possible hazards to human health. This axiom is founded on the quantal dose–response concept that the occurrence of a toxic response in a population increases as the dose and duration of exposure increases. Practical limitations in the design of toxicological studies dictate that a small number of animals will be exposed to the chemical compared with the size of the human population potentially at risk. Consequently, relatively large doses of chemicals are necessary to increase the frequency with which treatment effects occur in small groups of animals. For example, suppose the incidence rate of a particular chemical to produce a tumor in humans was as low as 0.01%. This incidence rate represents 24,000 people in a population of 240 million people. A toxicological study would have to use a minimum of 30,000 animals in order to detect a tumor in three animals! It is not feasible to use exceedingly large numbers of animals to assess the toxic properties of chemicals. The only available option is to collect toxicity data from small groups of animals exposed to large doses of chemicals and to use the principles of toxicology and pharmacokinetics to extrapolate the experimental findings to humans. In this manner, toxicologists can assess human health hazards that result from low-dose exposures to environmental pollutants and other chemicals found in the workplace and home.

Toxicity Studies. Test species, number of animals, duration of chemical exposure, dose level, and the measured parameters of toxicity are the fundamental variables that constitute all descriptive toxicological studies.

Animals. Most often, the selection of species in toxicity studies ultimately depends on the regulatory agency that will review the data. Regulatory bodies, in turn, set species requirements based on many years of assessing safety data submitted by industry as well as the advice from recognized authorities in the field of chemical hazard assessment. The most important criterion for species selection is that the spectrum of toxicities observed in humans be reproducible in the test species. This is recognized retrospectively for a plethora of compounds that entered the market or environment before the advent of rigorous toxicity testing. These toxicities preferably are elicited at dosages equivalent to, or lower than, those encountered by humans and expressed after relatively short durations of exposure.

The EPA usually requires rats and mice to be used as test species in studies that characterize the toxic effects of pesticides and other toxic substances that may cause occupational injury or contaminate the environment. Exceptions to this rule occur when a specific adverse effect seen in humans is not easily reproduced in rodents. For instance, some organophosphorus compounds, such as triorthoscresyl phosphate, produce a neurotoxic effect in humans manifested by a "dying back" of the distal ends of long, large-diameter, peripheral nerves. The chicken is highly sensitive to this lesion and is used routinely to determine whether a new pesticide has this toxicological property.

The FDA typically requires a rodent and a nonrodent species for the assessment of drug safety in subchronic and chronic studies that determine which organs are affected by high-dose drug treatment. The rat is the rodent species used most often, whereas the nonrodent species is either the beagle dog or the cynomolgus monkey. Acute toxicity tests (i.e., single-dose administration) evaluate the possible consequences of drug overdose and employ three test species, usually the mouse, rat, and rabbit. Rats and rabbits are used to evaluate the effects of chemicals on male and female fecundity as well as fetal development. Mice and rats are used to determine whether a compound produces tumors (i.e., carcinogenicity study) when administered throughout an animal's entire life span (e.g., 1.5–2 years for mice; 2–2.5 years for rats). In addition, most studies that evaluate the systemic toxicity of chemicals contain both male and female animals.

Numbers of Animals. The number of animals employed in regulated toxicity studies that support the marketing approval of a drug, food additive, or agricultural chemical is defined by the regulatory agency that has jurisdiction over the indicated usage of the chemical. The minimum number can be as low as a single animal as in the case with acute toxicity tests that employ dogs as the test species. For example, the lethal dose in dogs is approximated by increasing the dose given to the same animal each day until serious side effects of the drug are observed. The dog is monitored for signs of recovery, removed from the study, and returned to the dog colony. When rodents are used in acute toxicity tests, as many as 10 animals per sex are assigned to a dose level.

In subacute studies, the test material is usually administered for 14 days. These studies provide an initial identification of target organs and are used to set dose levels for subchronic studies, that is, studies of 1 and 3 months in duration. Subacute studies usually are not regulated, because they are not used directly for human hazard assessment. Accordingly, the descriptive toxicologist and not a governmental guideline selects the number of rats and dogs to be used in subacute testing. Frequently, sample sizes for rats and dogs are 5–10 per sex and 2–4 per sex, respectively. In contrast, repeat-dose, target organ studies submitted to regulatory bodies (i.e., subchronic and chronic studies) are required to have 10–25 rodents per sex per dose level and 3–6 dogs or monkeys per sex per dose level.

In carcinogenicity studies, as many as 50–75 rodents per sex per dose level are placed on diets that contain the test substance. Study designs include at least three treatment groups on diets that have increasing concentrations of test material and often call for two groups to be placed on control diets. Thus, these lifetime studies frequently begin with 500 or more rodents. Large numbers of animals are needed to discriminate between treatment-related increases in tumor incidence and those that occur spontaneously in senescent animals. Accordingly, regulatory agencies may request that a minimum of 30 animals per sex per dose level survive to the end of the study. For reasons that remain obscure, the 24-month survival rate of naive rats employed in toxicity studies has decreased somewhat over the years. Consequently, large numbers of animals are placed on study to ensure that the 30-animal requirement is met.

Route of Administration and Dose Levels. The route of administration employed in toxicity studies is the same as that expected for human exposure and is often oral. Oral administration is achieved by direct placement of the test material in the stomach (i.e., oral gavage) or by mixing the compound into the diet, which is then fed to the animal. Toxicity studies employ a minimum of three dose levels and a single control group. Ideally, the high dose should clearly demonstrate some form of a noxious response. The lowest dose level should identify the dose that produces no observable adverse effects (NOAEL), whereas the mid-dose level is included to establish a dose–response relationship and determine the lowest dose that produces adverse effects (LOAEL).

Parameters of Toxicity. The parameters recorded in subacute and subchronic studies include clinical observations (i.e., body weight gain, food consumption, physical appearance, clinical behavior, cardiovascular and respiratory distress, loose stools, and macroscopic appearance of the eyes), clinical chemistry (i.e., serum levels or activities of sodium, potassium, calcium, chloride, glucose, blood urea nitrogen, creatinine, bilirubin, alanine aminotransferase, aspartate aminotransferase, sorbitol dehydrogenase, alkaline phosphatase, γ-glutamyl transpeptidase, triglycerides, albumin, and total protein), hematology (i.e., assessment of blood for hemoglobin, hematocrit, red blood cell count, white blood cell count and their subtypes, platelet count, and clotting time), urinalysis (i.e., examination of urine for color, volume, specific gravity, pH, blood, glucose, ketones, protein, bilirubin, and urobilinogen and the presence of crystals, casts, epithelial cells, red blood cells, white blood cells, and microorganisms), gross and light microscopic appearance of all major tissues (i.e., brain, heart, liver, kidney, spleen, testes, prostate, epididymus, seminal vesicles, ovaries, uterus, thyroid, parathyroid, adrenals, thymus, eyes, lungs, lymph nodes, mammary glands, spinal cord, sciatic nerve, pituitary gland, pancreas, bone, bone marrow, skin, skeletal muscle, and various portions of the urinary and gastrointestinal tracts), and the weight of the first 15 aforementioned tissues.

Chronic and Lifetime Carcinogenicity Studies. Chronic studies and lifetime carcinogenicity studies are used to assess adverse reactions that require protracted exposure times. The study design and objectives of the chronic study are similar to those of the subchronic study. The duration of the chronic studies depends on the length of human exposure envisioned for the test material as well as the test species. For example, a 6-month chronic study is appropriate to assess the chronic target organ toxicity of an antimicrobial agent having an anticipated clinical treatment regimen of 10 days in duration. In general, the period of exposure in chronic rodent studies is between 6 and 12 months. Dogs and monkeys typically receive test materials for 1 year. However, much longer exposures are occasionally employed, as was the case with oral contraceptives that were administered to monkeys for several years. Chronic studies frequently incorporate recovery groups that are monitored for several months or more after the cessation of compound administration to determine whether the toxicities noted during treatment are reversible.

The lifetime carcinogenicity study is colloquially referred to as a "bioassay" and is utilized to determine whether a compound has the ability to produce tumors. Dose selection is critical for carcinogenicity studies, because marked chronic intoxication, morbidity, and mortality will reduce the number of animals that survive to the term of the study, which has a serious impact on the acceptability of the study by foreign and domestic regulatory agencies. Thus, a maximum tolerated dose (MTD) is used as the highest dose level in carcinogenicity studies, and fractions thereof are used as mid- and low-dose levels (e.g., one-half and one-quarter of the MTD). The MTD is frequently defined in the pharmaceutical industry as a dose that suppresses body weight gain by a maximum of 10% and is often identified in the 3-month subchronic study. A strong commitment to bringing a product to market is made when a pharmaceutical, food, or agricultural chemical company decides to conduct carcinogenicity studies, because they are animal-, test material-, time-, and labor-intensive and are by far the most expensive toxicity tests to conduct. The current cost of conducting carcinogenicity studies in rats and mice is approximately $2 million.

Summary

In summary, the principles of toxicology are employed to design sensitive and predictive animal safety studies with the ultimate goal of extrapolating the data to human and environmental chemical exposure situations. In this manner, the hazards that chemicals pose to the health of humans and the well-being of the environment can be assessed and, more important, averted.

RECEIVED for review September 3, 1992. ACCEPTED revised manuscript February 11, 1993.

Toxicology and Risk Assessment

Arthur L. Craigmill[1], Scott Wetzlich[1], and William M. Draper[2]

[1]Environmental Toxicology Extension, University of California, Davis, CA 95616
[2]Hazardous Materials Laboratory, California Department of Health Services, 2151 Berkeley Way, Berkeley, CA 94704

> *Toxicology and epidemiology are inextricably linked in recognizing and regulating chemical hazards. Although epidemiological studies are widely appreciated as the "gold standard" in human health risk assessment, for ethical reasons it is often impossible or unacceptable to base risk management decisions on human epidemiological data alone. The toxicological sciences identify potential hazards to human health by using experimental animal models, which act as surrogates to humans. This review provides an overview of toxicological risk assessment and its four principal components: hazard identification, dose–response assessment, exposure assessment, and risk characterization. The contrast between risk characterization for threshold toxicants and that for nonthreshold toxicants, a major distinction in regulatory toxicology, is considered. Reliance on new and more sophisticated scientific tools at each stage of risk assessment continues to strengthen the scientific basis of the process.*

THE POPULARITY OF RISK ASSESSMENT was predicted by John Frawley in the summer of 1980 when he said that risk assessment would become "the fad of the eighties" (1). The popularity of risk assessment was derived from its ability to provide answers to questions about chemical exposure. Frawley's prediction was accurate, except that his statement seems to be true of the 1990s as well. Consistent with this enduring interest, numerous excellent review articles and volumes have been written over the past 15 years, and a short list of recent examples is provided (2–8).

Toxicology and epidemiology are inextricably linked in recognizing and regulating chemical hazards. Epidemiologists often consider toxicological information in formulating hypotheses, which are then tested by observation in human populations. In addition, animal toxicological studies may complement and amplify the results of epidemiological studies by revealing biochemical modes of action and thereby establishing plausible disease etiologies. Toxicologists, in contrast, recognize the preeminence of well-conducted, human epidemiological studies, which provide direct evidence of an association between a chemical (or other factor) and human disease without extrapolating between species and across tremendous dose ranges.

This chapter does not attempt to provide a comprehensive or in-depth review of current risk-assessment science. Rather, it provides an introductory overview that complements our treatment of environmental epidemiology. An appreciation of the strengths and limitations of toxicological risk assessment may help us better appreciate the role of human studies. Risk assessment, like any other scientific endeavor, is continually undergoing refinement. Therefore, current controversies are discussed briefly; undoubtedly, these areas will be better understood by future generations.

Toxicological Risk Assessment

The prediction of the type of toxicological effect caused by exposure to a chemical, the magnitude of the effect, and the incidence of the effect within an exposed population are all aspects of toxicological risk assessment. The assessment of risk from chemical exposure requires information from an extensive array of toxicological tests but may also rely on epidemiological studies. These toxicity tests include acute and chronic exposures in whole animals and short-term in vitro tests utilizing bacteria, animal cells, and subcellular enzyme preparations. Human epidemiological data, when available, provide the most relevant information. In the absence of human studies, risk assessments rely on data from experimental animal models; these data are extrapolated to indicate both the type of effects possible in humans and the relationship of dose to toxic response.

The National Academy of Sciences' paradigm of chemical risk assessment identifies four stages: (1) hazard identification, (2) dose–response assessment, (3) exposure assessment, and (4) risk characterization (9). The prediction of risk usually distinguishes two broad categories, depending on whether toxicity is considered to be a threshold or nonthreshold phenomenon. Threshold toxic effects are not measurable until the dose exceeds a threshold value—all exposures below this level are believed to be nondeleterious. Only mutagens and carcinogens are currently treated as nonthreshold toxicants.

Risk assessments are usually performed to avoid unacceptably high chemical exposures (proactive) or to predict the possible incidence of toxic effects in a population after exposure. Risk assessments do not provide actuarial estimates of chemical-induced illness or death; rather, they provide estimates and confidence limits of population-wide risk as extrapolated from experimental studies or human epidemiological data.

Regulations based on risk assessment, therefore, may play an important role in protecting public health, the principal goal of which is primary disease prevention. The validity of risk assessments (and thus the cost-effectiveness of health-based environmental regulations) are only as good as the data on which they are founded.

Hazard Identification. The first task faced by risk assessors is to establish whether a given compound poses a hazard to human health. Although this determination appears to be qualitative, dose cannot be ignored, because it is well known that every compound is toxic at some level. Therefore, the pertinent question is whether the compound may be harmful to humans over the range of exposures likely to be experienced by the population in environmental or occupational settings.

The analytical process involved in hazard identification represents a preliminary evaluation of all available toxicological, epidemiological, and exposure information. It establishes the rationale and urgency for further investigation. As in all scientific investigations, the truth is inferred from information derived from both experiments (toxicological testing) and observation (epidemiological studies). The role of scientific inference and the pressure on risk assessors to "see" the data from a particular point of view make risk assessments inherently controversial.

The types of information central to hazard identification include human data and laboratory data. Human data (i.e., from occupational exposures, accidents, and suicides) provide an indication of the degree to which a chemical may be hazardous to a broader population. However, this information is often complicated by a lack of exposure quantification. Laboratory data from experimental animals provide reliable exposure information at the cost of uncertainty due to extrapolation between species.

Commercial chemicals to which humans may be exposed (e.g., pharmaceuticals, food additives, and pesticides) are tested extensively before their approval for general use—far more so than industrial chemicals, intermediates, and chemical constituents of consumer products. Toxicity testing is conducted according to a rational, tiered approach. In this manner, the most costly and elaborate tests are postponed pending the outcome of screening tests. In the development of a chemical for market, unsatisfactory results from preliminary toxicological testing may eliminate a candidate compound from further consideration.

Structure–Activity Relationships. Comparing the structure of a new chemical to that of other compounds with known biological activity is the basis of a structure–activity relationship (SAR) analysis. Just as the desired biological attributes of chemicals (e.g., pesticidal action against insects or pharmacological activity at a particular receptor) follow relationships predictable by chemical structure, so do a chemical's undesirable toxicological side effects. SAR testing is rapid and relatively inexpensive, and commercial computer software can evaluate acute toxicity, skin and eye irritation, mutagenicity, and carcinogenicity (10).

In Vitro Toxicity Tests. A battery of in vitro tests typically follows. In vitro tests are advantageous, because they do not involve whole animals but, rather, single-celled organisms, cell cultures, or subcellular systems (e.g., enzyme preparations) and reveal biological activity associated with toxicity in vivo. A familiar example is use of acetylcholinesterase enzymes of red blood cells as a probe for mammalian neurotoxins. Another well-known in vitro test is the bacterial mutagenesis assay, also known as the *Salmonella* or Ames mutagenicity test, which detects animal carcinogens as bacterial mutagens.

Although a positive finding in any of the in vitro tests attracts the attention of toxicologists, an isolated positive result among a series of negative findings is not uncommon. The in vitro tests provide the optimal conditions for expression of activity. In the intact organism, the biological activity of a chemical may be mitigated by barriers to absorption, by metabolism and conjugation that promote detoxication and excretion, and by other processes. Furthermore, the peculiarities of the bioassay may lead to exceptional biological activity that has no corresponding relationship to biological activity in intact higher organisms. *Salmonella* bacteria, for example, are exquisitely sensitive to mutation by some nitrated polycyclic aromatic hydrocarbons (PAH) without activation by microsomal enzymes, whereas under similar test conditions, known human carcinogens like benzo(*a*)pyrene produce a negative result. Thus, in vitro tests are only suggestive of toxicity.

The array of in vitro tests for detection of carcinogens, mutagens, and reproductive toxicants are reviewed in this volume in Chapters 7 and 8.

Toxicity Testing in Laboratory Animals. The next tier in toxicity testing involves experimental animals, or in vivo testing, which provides information on toxicity to whole animals and preliminary information on the doses required to elicit toxic effects. The spectrum of tests commonly utilized has been outlined in Chapter 5. These tests are conducted to evaluate a variety of toxicity end points, and they rely on the study of multiple species, dose levels, and exposure periods. Acute exposure tests are typically followed by repeated dose studies and ultimately by chronic studies for reproductive toxicity and carcinogenicity. A more complete list of laboratory toxicity tests is presented in List I. Much of the toxicity data originating in the hazard identification phase is useful at the next stage, dose–response assessment.

Dose–Response Assessment. Two pieces of toxicological information are critical in health risk assessment: the existence of a threshold and the nature of the dose–response relationship. Detailed knowledge of the

List I. Toxicity Tests Used in Toxicological Risk Assessment

Structure–activity relationships
  Physicochemical properties
  Similarity to known toxicants
In vitro test results
  Mutagenicity
  Other
In vivo test results
  Short-term (acute)
    Lethality
    Organ toxicity
    Irritation
    Sensitization
    Genotoxicity
    Repeated dose
  Chronic
    Reproductive
    Carcinogenicity

dose–response relationship may reveal a threshold and thus the two aspects are closely related. As in other stages of risk assessment, information from human exposures is the most relevant and the most sought-after. Information from occupational exposure studies are sometimes available, but a major difficulty—especially with retrospective studies—is misclassification. Misclassification of exposure groups tends to bias risk estimates toward the null; in other words, it leads to an underestimate of the hazard. Accordingly, accurate exposure classification is a fundamental requirement for any well-designed study. For new chemicals, human exposure data are not available, which necessitates reliance on laboratory test data from experimental animals. A partial list of laboratory studies used in establishing dose–response relationships is shown in List II, and a number of authoritative review articles on this topic are cited in the National Academy of Science's report (9).

Exposure Assessment. Estimation or measurement of chemical exposure is essential to estimating dose, which, in turn, determines the range

List II. Toxicology Tests Used in Defining Dose–Response Relationships

Acute lethality	Immunotoxicity
Skin and eye irritation	Hematotoxicity
Central nervous system toxicity	Reproductive toxicity
Peripheral neuropathy	Teratogenicity
Gastrointestinal toxicity	Mutagenicity (in vivo)
Respiratory toxicity	Carcinogenicity
Hepatotoxicity	Behavioral toxicity
Nephrotoxicity	

and severity of toxic effects in the exposed population. Absorption is defined as the movement of a chemical from the site of administration into the blood. Toxicity is determined by the concentration of the active form of the chemical (which may be a metabolite) at the site of action or cellular receptor. This information is not directly available.

Information on the internal or absorbed dose is most pertinent for risk assessment. This dose can sometimes be estimated from the concentration of a chemical, or that of a metabolite, in body fluids. Like toxicity testing, however, there is a battery of dose estimation techniques. Where human biomonitoring data are unavailable, environmental or personal exposure monitoring data may establish exposure concentrations, residue concentrations in the environmental compartments in immediate contact with the study population. Exposure concentrations can also be estimated by using ad hoc methods or modeling.

Biological Monitoring. Human biological monitoring provides direct evidence of exposure as described in several of the chapters in this volume, particularly Chapters 9 and 10. Chemists and biochemists have made significant advancements in the technology for measurement of chemical residues in biological systems. For an up-to-date overview of human biomonitoring technology, readers are referred to the proceedings of a recent biomonitoring symposium (*11*). Biological monitoring encompasses more than measurement of internal dose, however, and has the potential for probing further into characteristics of the exposed population (e.g., susceptibility, effects, preclinical disease), as reviewed in Chapter 9 and previously by Perera (*12*). This is currently a very active area of research.

Human tissues are sometimes available from poisoning victims, and samples from live subjects may include bodily fluids (e.g., blood, milk, or saliva), exhaled breath, hair and nails, skin wipes, urine, and feces. Human biopsy samples also are occasionally available where laboratory researchers have developed collaborations with hospitals. Urine is the most easily obtained biological matrix, whereas blood samples must be obtained by a phlebotomist or other trained health professional.

Although human biomonitoring is recognized as the most reliable measure of exposure, it is not a panacea. First, the data are often not available, because the analytical techniques—especially for metabolites (the main chemical excretion products)—are restricted to a few research laboratories. There are no government-certified laboratories or analysis procedures for biomonitoring, and commercial environmental testing laboratories do not routinely offer biological monitoring services. Furthermore, there is little demand for such procedures, because this type of testing is not mandated by any federal or state laws or regulations. Laboratories that have expertise in industrial hygiene techniques typically offer personal monitoring, which involves the measurement of chemicals in the breathing zone. Second, biological monitoring data do not lead directly to an estimate of dose, as pointed out in Chapter 14. As the chapter demonstrates, pharmacologically based, pharmacokinetic models represent a powerful interpretive tool when applied to biological monitoring data.

Measurement of Exposure Concentrations. When human biomonitoring data are absent, which is usually the case, chemical analysis of environmental media can provide satisfactory estimates of human exposure in particular situations. Data on chemical residues in the environmental compartments, so-called exposure concentrations, are combined with estimates of food, water, and air intake to approximate exposure by the relevant pathways.

Theoretical Models in Exposure Assessment. In many instances, even environmental monitoring data are not available. As a last resort, exposure estimates can be based solely on modeling or ad hoc methods, and this is often necessary after industrial and transportation accidents and spills. Chapter 15 provides an example of exposure estimation based on environmental modeling. In that study, a transportation spill of the fumigant metam-sodium, little or no reliable environmental data were available on the atmospheric concentrations of the irritant breakdown product, methyl isothiocyanate.

Because a chemical released to the environment partitions into various environmental compartments, including soil (lithosphere), air (atmosphere), water (hydrosphere), and biological organisms as well as the associated subcompartments (e.g., vadose zone, atmospheric particulate matter), modeling exposures by each pathway can be very complex. The phrase "multimedia exposure assessment", popular today, acknowledges the diversity of exposure routes to humans and the complexity of evaluating the fate of a chemical in the environment. Clearly, wherever possible, exposure assessments based on theoretical models need to be validated (or calibrated or confirmed) by bioenvironmental or environmental monitoring or by other approaches, such as laboratory simulation. Moreover, the uncertainties associated with these exposure estimates need to be provided to public health authorities relying on the information.

Risk Characterization. Risk characterization is the process of extrapolating dose–response data from human and laboratory animal studies to probable exposure scenarios to predict the incidence and degree of toxicity in the exposed human population.

Safety Factors for Threshold Toxicants. In risk characterization, a major distinction is made between threshold and nonthreshold toxicants. For risk assessment purposes, exposures below a threshold are believed to be without adverse effect, at least for the toxicological end point considered.

By applying a further safety factor, usually from 100 to 500, an acceptable margin of safety can be ensured. If a pollution standard is set at a level that protects a sensitive subpopulation (e.g., air pollution standards set below the threshold for asthma sufferers or the elderly), there is a high degree of confidence that no adverse effects will be experienced by the general population.

Johannsen has reviewed the historical development, use, and rationale for safety factors in risk assessment and regulatory toxicology (13). Reliance on safety factors addresses a number of limitations in toxicological risk assessment, including inadequacies of the data, interspecies adjustment, and intraspecies variability. Intraspecies variability is attributed to the heterogeneity found in the human population. Furthermore, differences among species are related to a complex interrelationship of multiple factors (genetic, physiological, and anatomical), which necessitate scaling. There are various approaches to scaling: Equivalent dose can be derived from body weight, body weight raised to an exponent, or surface area, or it can be derived allometrically.

The World Health Organization (WHO) established the acceptable daily intake (ADI) to define the permissible intake for chemicals that exhibit toxicity thresholds. The ADI is defined as "the daily intake of a chemical which, during an entire lifetime, appears to be without appreciable risk on the basis of all known facts at the time" (14). The U.S. Environmental Protection Agency has established a similar criterion called the Reference Dose (RfD).

The origin of both the ADI and the RfD is the no observable adverse effect level (NOAEL) determined in the experimental animal model, which is divided by a safety or uncertainty factor. The ADI and the RfD may differ, because different factors may be employed (15) or different toxicity end points may be considered.

The threshold limit value (TLV) is another example of a safe-exposure guideline established for compounds that have toxicity thresholds. The American Conference of Governmental Industrial Hygienists (ACGIH) defines the TLV as "the concentration for a normal 8-hour workday and a 40-hour workweek, to which nearly all workers may be repeatedly exposed, day after day, without adverse effect" (16).

Quantitative Risk Assessment for Nonthreshold Toxicants. For regulatory toxicological purposes, mutagenicity and carcinogenicity are considered to be nonthreshold events, because a single chemical interaction with the genetic material may, in theory, induce disease. Therefore, to predict cancer risk requires extrapolation along a theoretical dose–response curve to the low-dose region, usually far removed from the dose range for which animal data exist. In a recent review, Johannsen describes several mathematical models that have been developed to extrapolate into low-dose regions (*13*). All of these models appear to be reliable for the range in which the data lie. At low doses, however, the obtained risk estimates vary over several orders of magnitude. In the United States, regulatory agencies rely on the linearized multistage model, a mathematical risk characterization model that assumes (1) that dose–response relationships are linear and (2) that cancer induction is a multistage process.

The development of biologically or physiologically based models of risk assessment is being pursued to help overcome the uncertainties of current theoretical models (*17*). Biologically based models consider in much greater detail the diversity of available chemical and biochemical information on mechanisms of action, metabolic activation and detoxication, and repair processes, which may be critical in mitigating toxicity observed at low doses. Moreover, as demonstrated in Chapters 11 and 12, species have different responses to chemical insults, and it is possible, by using bioanalytical chemistry, to distinguish sensitive and insensitive species.

The disparity in regulations and exposure guidelines for threshold and nonthreshold toxicants is exemplified in the recommendations for carcinogens made by the American Conference of Governmental Industrial Hygienists (ACGIH) (*16*). Workers handling confirmed human carcinogens "should be properly equipped to eliminate to the fullest extent possible all exposure". Some TLVs have been established for carcinogenic substances, but in these cases the exposures are not likely to result in a detectable increase in cancer incidence or mortality.

Uncertainties in Risk Assessment

There are numerous sources of uncertainty in predicting human health risks based solely on animal test data. Risk assessments that incorporate epidemiological study data, of course, are not without uncertainty, and valid experimental data from human studies also require a "safety factor" when extrapolated to humans in other exposure situations (*13*).

All scientific information, including that from laboratory toxicology studies, is accompanied by error. Experimentalists must continually track and mitigate both random and determinate errors, even when the design of their experiments is completely adequate. Toxicologists, like other biologists, must consider the diversity of biological populations, although laboratory animals are generally highly inbred. Accordingly, attention to statistics, appropriate sample sizes, and adequate experimental design are critical in toxicology.

There are other methodological problems. The high doses needed to produce observable effects in a limited number of experimental animals dosed for a limited period of time are often orders of magnitude greater than potential human exposures. When this is the case, the upper and lower confidence intervals "flare out" as the data are extrapolated farther and farther from the experimental doses. In some cases, the upper and lower bounds differ by four to six orders of magnitude.

In carcinogenicity bioassays in the United States (but not Europe) (18), two animal species (usually mice and rats) are exposed to a substance for 2 years at the maximum tolerated dose (MTD). At least one other dose group (usually 0.5 × MTD) and a control group are also maintained. The MTD is the amount of substance that causes some toxicity (e.g., a change in weight gain) during a 90-day exposure but does not detectably shorten the animal's life span.

Chemical toxicokinetics at such levels may be considerably different from those at low exposure levels, and certain detoxication pathways may be overwhelmed. At very high doses, cell damage and death may induce cellular proliferation, which itself may promote tumor formation (19). Thus, because the tumor-induction mechanisms are unique, the relevance of carcinogenicity testing at the MTD is controversial to some scientists. Lu and Sielken (20) postulate that the model used in current regulatory risk assessment "is really not a measure of low-dose carcinogenicity, but rather an indication that at toxic doses a chemical is also carcinogenic". A rapid resolution of the scientific question of whether carcinogens have thresholds is unlikely.

One source of uncertainty in risk characterization is the extrapolation from the laboratory to the real world and from experimental animals to humans. Risk assessors are currently using new physiologically based pharmacokinetic (PBPK) models to help reduce this uncertainty. PBPK models relate human exposure to laboratory animal exposure by comparing organ blood flow and toxicant extraction ratios between species. These data are incorporated into multiple, simultaneous differential equations that estimate more precisely the distribution of a chemical to target organs (21,22).

These and other sources of uncertainty in toxicological risk assessment elevate the significance of human data. Even when human data are available, however, all that may be known is that a chemical causes different effects in humans or that humans are more or less sensitive than experimental animals. Good human dose–response data for environmental chemicals are usually lacking. The most abundant data available are for human pharmaceutical preparations used to treat chronic diseases, for example, anticonvulsants and tranquilizers. The difficulty in establishing exposures is a recurring problem for retrospective human epidemiological studies. Prospective human epidemiological studies, particularly of occupational exposures, offer some of the most relevant information available (23).

Summary

Toxicological risk assessment is characterized by substantial technical challenges, areas of uncertainty, and, frequently, heated controversy. For ethical reasons, most of the findings and conclusions derived from toxicological risk assessment cannot be tested experimentally. In spite of these limitations, the process is essential in protecting public health and the environment.

New models that probe biological mechanisms of action and that can account for differences in toxicokinetics among species will help strengthen the scientific basis of risk assessment and the validity of risk management decisions. Even with these and other advancements, toxicological risk assessment is not expected to provide actuarial data of the sort provided by human epidemiological studies. Human epidemiological studies relying on valid exposure classification techniques, including biomonitoring, will continue to provide the strongest evidence of health hazards for population-wide exposures to environmental chemicals.

For ethical reasons, it is often impossible or unacceptable to base risk management decisions on human studies. Many carcinogens have long latency periods, and in large populations, even low-level exposures to weak carcinogens may result in unacceptably high mortality and morbidity rates. Hence, the urgent need for toxicological risk assessment (and utilization of the best available scientific information in the process) is likely to remain with us for the foreseeable future.

It has been stated that risk assessment is fundamentally a social, cultural, and political phenomenon that has only incidental connection to science. Information based on scientific data, however, has always formed the most satisfactory and ultimately acceptable basis for resolving technical problems. Accordingly, improved integration of information gleaned in the toxicological laboratory and that obtained from well-designed epidemiological studies will lead to reliable risk assessments. This in turn can lead to regulations that justify their economic cost by providing the necessary level of protection for public health and the environment.

References

1. Frawley, J. J. *Reg. Toxicol. Pharmacol.* **1981**, *1*, 3–7.
2. Krewski, D.; Murdoch, D.; Withey, J. R. *Health Phys.* **1989**, *57* (Suppl. 1), 313–325.
3. Krewski, D.; Wigle, D.; Clayson, D. B.; Howe, G. R. *Rec. Results Cancer Res.* **1990**, *120*, 1–24.
4. Beck, B. D.; Toole, A. P.; Callahan, B. G.; Siddhanti, S. K. *Reg. Toxicol. Pharmacol.* **1991**, *14*, 273–285.
5. Koehman, J. H. *Comp. Biochem. Physiol. C. Comp. Pharmacol. Toxicol.* **1991**, *100*, 7–10.
6. Leung, H. W. *J. Toxicol. Environ. Health* **1991**, *32*, 247–267.
7. Rosati, G.; Morisetti, A.; Tirone, P. *Toxicol. Lett.* **1992**, *64–65*, 705–715.
8. Scheuplein, R. J. *Crit. Rev. Food Sci. Nutr.* **1992**, *32*, 105–121.
9. *Risk Assessment in the Federal Government: Managing the Process*; National Research Council; National Academy Press: Washington, DC, 1983.
10. Enslein, K. *Toxicol. Ind. Health* **1988**, *4*, 479–498.

11. *Applications of Molecular Biology in Environmental Chemistry*; Minear, R.; Karch, N. J.; Needham, L. L.; Ford, A. M., Eds.; Lewis: Chelsea, MI, in press.
12. Perera, F. P. In *Biological Markers in Risk Assessment;* Travis, C. C., Ed.; Plenum: New York, 1988; pp 123–138.
13. Johannsen, F. R. *Crit. Rev. Toxicol.* **1990**, *20*, 341–367.
14. World Health Organization. *Evaluation of the Toxicity of a Number of Antimicrobials and Antioxidants*; WHO Technical Report Series; World Health Organization: Geneva, Switzerland, 1962; p 228.
15. Lee, S. D. *Toxicol. Ind. Health* **1990**, *6*, 245–255.
16. American Conference of Governmental Industrial Hygienists. *Threshold Limit Values for Chemical Substances and Physical Agents and Biological Exposure Indices;* American Conference of Governmental Industrial Hygienists: Cincinnati, OH, 1990.
17. Portier, C. J.; Hoel, D. G.; Kaplan, N. L.; Kopp, A. In *Complex Mixtures and Cancer Risk;* Vanio, H.; Sorsa, M.; McMichael, A. J., Eds.; International Agency for Research on Cancer: Lyon, France, 1990; pp 20–28.
18. Chan, P. K. *J. Toxicol. Sci.* **1990**, *15* (Suppl. 4), 176–182.
19. Ames, B. N.; Gold, L. S. *Prog. Clin. Biol. Res.* **1991**, *369*, 1–20.
20. Lu, F. C.; Sielken, R. L. *Toxicol. Lett.* **1991**, *59*, 5–40.
21. Grant, L. D.; Jarabeck, A. M. *Toxicol. Ind. Health* **1990**, *6*, 217–233.
22. Travis, C. C. *Pharmacokinetics and Carcinogen Risk Assessment;* Travis, C. C., Ed.; Plenum: New York, 1988; pp 87–102.
23. Krewski, D.; Wigle, D.; Clayson, D. B.; Howe, G. R. *Rec. Results Cancer Res.* **1990**, *120*, 1–24.

RECEIVED for review September 3, 1992. ACCEPTED revised manuscript June 15, 1993.

7

In Vitro and In Vivo Assays to Screen for Reproductive Toxicants in Animals and Humans

Barbara S. Shane

Institute for Environmental Studies, Louisiana State University, Baton Rouge, LA 70803

Although more than 800 compounds that have been tested in animals have been shown to be embryotoxic or teratogenic, little is known of the reproductive toxicity of these and other chemicals in humans. Because the compounds that have been tested account for only 2.5% of the potentially embryotoxic or teratogenic compounds to which humans are exposed, emphasis has been placed on the development of short-term reproductive assays to screen these compounds more rapidly. In some cases, the results obtained in the in vivo assays in intact animals correlate well with those obtained in the short-term in vitro assays; however, the usefulness of these assays will be evident only after extensive validation. Also, epidemiological studies in humans are difficult to interpret when the cohorts are small and the adverse reproductive outcomes very subtle. In this chapter, the most commonly used in vitro and in vivo assays are described in association with the outcomes elicited by certain classes of chemicals. Additionally, epidemiological studies in humans in which adverse reproductive outcomes have been elicited by industrial chemicals are also discussed.

THE HUMAN POPULATION HAS BEEN EXPOSED (*1*) to approximately 60,000 chemicals, and this number is increasing by a few thousand annually because of the synthesis of new compounds, inadvertent production in industrial processes, and chemical interactions in the environment (*2*). Of these chemicals, only 1600 have been evaluated for their fetotoxic and teratogenic potential in animals. Close to one-half of these compounds have been shown to be teratogenic in animals, whereas less than 50 compounds have been shown to be teratogenic (1) or cause spontaneous abortion in humans (3–5). The number of categories into which these chemicals fall is numerous and includes pesticides, industrial compounds, metals, organic solvents, laboratory chemicals, drugs, feed additives, environmental chemicals, and naturally occurring substances such as plant, fungal, and bacterial byproducts. The majority of these chemicals exist in the environment in negligible amounts; however, a small percentage have been shown to persist at toxic levels (2).

The toxic potential of chemical contamination of the environment presents hazards not only to the current population but also to future generations by adverse effects to the genetic material and reproduction. Exposure of both male and female organisms to chemicals during the reproductive cycle can have deleterious effects on the development of the fetus. The toxic action of environmental chemicals on reproduction occurs by a direct effect on the conceptus or by effects on maternal organs that can result in altered hormone secretion and hence altered reproduction (6). Estimating the incidence of these effects is difficult, and determining their causes is even more complex because of the large number of confounding factors. These factors include but are not limited to the following: incomplete information concerning dose, timing and duration of exposure; unknown interactions among causes; difficulty in obtaining specimens; difficulty in determining environmental concentrations; the large number of potential causes of abortion; and variations in individual susceptibility due to differences in genotype (7). To understand these effects on reproduction, this review addresses the physiology of pregnancy and fetal development, the etiology and mechanisms of reproductive failure and fetal malformations (teratogenesis), the evaluation of environmental pollutants for their developmental toxicity in in vitro and in vivo assays, and the assessment of the toxicity of these compounds in epidemiological studies.

Physiology of Pregnancy and Fetal Development

Gestation is the period of fetal development from the time of conception to the time of birth. The average length of gestation in man is 266 days, or approximately 9 months (8). Gestation is commonly divided into three periods each of 3 months duration, referred to as the first, second, and third trimesters.

The greatest damage to the developing embryo following exposure to toxic compounds occurs during the first trimester, when the embryo differentiates. The blastocyst, which is formed during the first week of gestation, consists of cells that have not undergone differentiation. Injury to a small number of these cells does not result in a specific developmental defect but in an overall delay in the development of the fetus or in death if the dose is high (9). Embryonic differentiation begins during the third week of gestation, when the cells of the blastocyst are segregated into three embryonic germ layers, referred to as ectoderm, mesoderm, and endoderm. During weeks 4 through 8, each of the three germ layers gives rise to a number of specific groups of cells, known as primordia. The differentiated cells have more specialized metabolic requirements and are, therefore, more vulnerable to damage by adverse influences. Damage to a particular primordium at this time may later result in specific structural defects in the fetus (Figure 1). On completion of organ formation (organogenesis) at the end of the twelfth week, the induction of major structural defects is no longer a factor of concern (9).

The period from the beginning of the third month to birth is known as the fetal period. It is characterized by rapid growth of the body and continued tissue differentiation or histogenesis (10). The progress of histogenesis is closely correlated with the development of the functional activity of the fetal organs. Adverse influences during this stage of development result in microscopic structural defects and possibly in functional abnormalities (9).

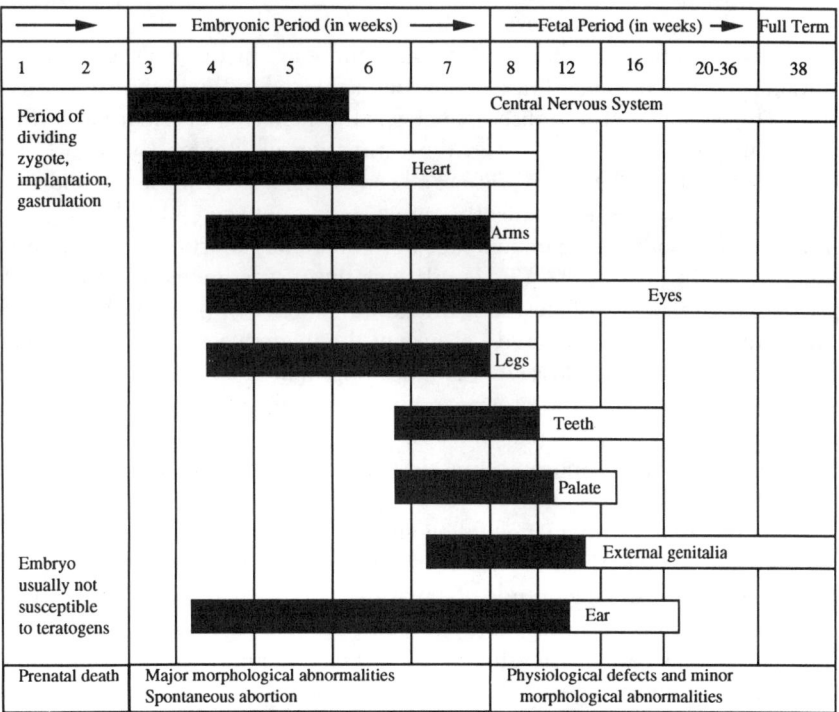

Figure 1. Human developmental stages and susceptibility of organ systems to reproductive toxicants. Dark hatched areas are time periods at which tissue is highly susceptible to environmental toxicants.

Because structural and functional maturation continues after birth in many organ systems such as the immunologic system, nervous system, liver, kidneys, and other endocrine organs, concern of the possible adverse effects of environmental factors on late-maturing functions during infancy and childhood is growing (9).

Reproductive Failure and Teratogenesis

Interference of normal development through exposure to various compounds may have one of four possible outcomes: death, malformation, growth retardation, or organ dysfunction of the fetus (9). Death of the embryo during the early stages of gestation (weeks 1 through 4) results in resorption of the conceptus by the maternal system, whereas death of the embryo during weeks 4 through 8 results in heavy bleeding that is frequently undetected as a fetal death by the mother. Many pregnancies are not diagnosed until the woman seeks medical attention, 8–10 weeks after conception. Death of the fetus during weeks 8 through 36 results in expulsion of the uterine contents, referred to as a spontaneous abortion. The rate (percentage) of spontaneous abortions is determined by multiplying the number of spontaneous abortions by 100 and dividing by the sum of the number of spontaneous abortions and the number of births in a specific population (11). The estimated rate of abortion reported in the literature is variable. The reason for these discrepancies is the inclusion or exclusion of the embryonic period (weeks 1–8) in the estimation of spontaneous abortion. One source estimates that 75–78% of all conceptions are resorbed or aborted (12), whereas another estimates the rate as being between 30 and 50% (13). Another study indicates the incidence of spontaneous abortion of recognized pregnancies to be 15–20% (3). Until recently, it was impossible to detect pregnancy in its early stages, and thus the incidence of pregnancy wastage was frequently underestimated because of the inadequacies of the methods of analysis of pregnancy loss (13). The development of a new and innovative immunoassay to detect the presence of urinary β chorionic gonadotropin, a hormone produced by the placenta 9 days after conception (14), has made it possible to determine more accurately the rate of spontaneous abortion in the first 2 months of pregnancy.

The outcome of pregnancy following exposure to a chemical substance depends on the length of exposure, the stage of fetal development at the time of exposure, the magnitude of exposure, and the nature of the chemical substance. Not all exposures result in fetal death and spontaneous abortion. Many exposures result in teratogenesis, which is defined as the production of any significant change in structure or function of the fetus that can be detected during the postnatal period (1). The possible consequences of teratogenesis are death of the fetus (spontaneous abortion), congenital malformation, growth retardation, and functional disorder of an organ system (15).

The final manifestation in the fetus depends primarily on the susceptibility of the embryo at the time of exposure and the magnitude of the dose (9).

The dose of the chemical that reaches the embryo or fetus depends on several factors: the magnitude of the dose, the physical form of the agent, the route of exposure, the rate of absorption by the maternal system, and the effectiveness of maternal metabolic processes that function to protect the fetus. The primary function of maternal metabolism is to reduce the blood concentration of the toxicant so that the number of molecules free to cross the placenta is minimized. This is accomplished by detoxification, excretion, or storage of the toxicant by the maternal system. The embryo or fetus is thought to have a threshold dose for most toxic substances. Below this dose, no effect occurs; above it, permanent changes in the fetus may be induced (9). Although the embryo or fetus may only receive a small fraction of the maternal dose, it is frequently sufficient to produce an embryotoxic or teratogenic response. Because of the extraordinary sensitivity of the conceptus, it is possible for embryotoxicity to occur in the absence of maternal toxicity following exposure (16).

Teratogens may alter normal development in a number of ways. These include gene mutations, chromosome breaks, interference with mitosis of cells, alteration in the normal function of the genes, lack of substrates required for growth through alteration in the source of energy, inhibition of certain enzymes, an osmotic fluid imbalance within the developing fetus, and changes in membrane characteristics. All of these cellular processes, which are fundamental to the development of an embryo, can be measured in vitro and thus can be studied separately. Although many teratogenic assays are relatively cheap and can be performed rapidly and cheaply, they model only a single developmental process. An alternative strategy being evaluated is the development of more complicated assays in which a number of developmental processes, including cell growth and division, cell-to-cell interaction and communication, and differentiation are measured simultaneously.

In Vitro and In Vivo Assays To Monitor for Reproductive Toxicants

Reproductive toxicity assays have been designed to detect embryotoxicity, embryolethality, and teratogenicity. Some assays may detect all three end points, but most detect only one. Three levels of screening are utilized in the evaluation of a compound for its effects on the developing fetus.

1. Prescreening employs simple, inexpensive, short-term testing and is most commonly performed in vitro (17,18).
2. Animal testing involves the testing of chemical agents for teratogenicity in small laboratory animals.

3. Epidemiological studies in humans attempt to show an association between a chemical and its reproductive effect (1).

In Vitro Short-Term Assays. In vitro assays can be performed by using cells, tissues, or intact organisms of various levels of complexity. Established cell lines assess the effect of chemicals on a single developmentally relevant end point of toxicity; primary cultures derived from embryos assess one or more of the developmental processes; cultures of organ primordia derived from embryos assess the development of a particular organ; and intact embryos, including rodent embryos in culture or free-living embryos of submammalian species, assess the toxicity of a compounds on a number of developmental processes. Many compounds examined in these assays have been shown to be embryolethal or to retard growth but not to be teratogenic. There is some confusion, however, as to whether the degree of damage is related to dose or whether some of the assays are biased toward detecting only embryolethal end points or growth-retardation effects. In fact, it has been stated that many of the in vitro tests are designed to detect embryolethality and not teratogenicity.

Some in vitro assays can be used to study the mechanism by which a compound elicits developmental toxicity. In assays using rodent embryo culture, the maternal factors that may contribute or alleviate toxicity are eliminated, as the embryos are removed from the mother and cultured in vitro. Because these embryos are unable to metabolize chemicals to active intermediates, the role of mammalian enzymes in the metabolism of the test compound can be evaluated. For example, it has been shown that a teratogenic effect is elicited by cyclophosphamide in rodent embryos in culture when adult metabolizing enzymes are included in the assay but not when these enzymes are omitted. Interestingly, only two metabolites of cyclophosphamide have been shown to be teratogenic (19, 20).

For any in vitro screen to be useful, it must have four attributes: predictivity, sensitivity, specificity, and elicitation of a quantitative response. It must be predictive of the toxicity of a compound in humans, and it must be quantitative, that is, there must be a dose–response relationship so that with increasing concentrations of the chemical, an increase in an effect is observed. Chemicals have a threshold at which they cause developmental toxicity. Because very few human epidemiological studies of the teratogenicity of chemicals have been undertaken, these screens have been used to predict teratogenicity in humans indirectly. Difficulties in extrapolation occur, because the dose levels required to elicit developmental toxicity in humans are largely unknown, except in the case of a few drugs such as thalidomide and methotrexate. A sensitive test is one that successfully identifies developmental toxicants with few false-negative results. A specific screen successfully classifies nontoxicants with very few false-positive results. False-negative or false-positive findings can be minimized by manipulation. For example, Steele et al. (*21*) and the National Toxicology Program (*22*) found that more false-negative results are generated when data from only the lowest concentrations of 44 substances, tested at concentrations of 0.5–40 mM, were evaluated. At a concentration of 0.05 mM, all the nonteratogens were correctly classified, giving an overall accuracy of 54%. As the concentration was increased, the number of false-negative findings decreased and the number of false-positive results increased. Overall, the accuracy was highest (71%) when the concentration was 10–20 mM. At the highest concentration, accuracy was lower and the number of false-positive results became extremely high, close to 90%. The accuracy of the battery was lowest at the concentrations at which false-positive and false-negative results were optimized.

None of the in vitro assays has been sufficiently validated or shown to be sufficiently predictive to be accepted as an all-purpose screen for developmental toxicants (Table I). However, one important application of rapid in vitro screens is in ranking chemicals within a family for developmental toxicity so that compounds with the highest potency can be studied in more depth in other systems. Four such studies were recently reported: one by Oglesby et al. (*23*) on the embryotoxicity of parasubstituted phenols in the postimplantation assay, a second by Mayura et al. (*24*) on chlorinated phenols using the *Hydra attenuata* and postimplantation rat embryo assays, a third by Kistler (*25*) on the teratogenic potential of a number of synthetic retinoids in the in vitro limb bud cell culture screen, and a fourth by Rawlings et al. (*26*), who ranked alkoxy acids (metabolites of glycol ethers) in the rodent postimplantation assay. Although the ranking system is informative, it is difficult to extrapolate from these data to the concentration that might be expected to elicit a toxic effect in embryos in vivo or in humans.

Table I. Predictability of Short-Term Tests

Assay	Predictability	No. of Compounds
Cell culture		
Mouse ovarian tumor	79	178
Human embryonic palatal mesenchyme	64	99
Neuroblastoma	86	57
Organ Culture		
Mouse limb bud	89	27
Whole Embryo Culture		
A. Invertebrates		
Hydra	100	24
B. Non-mammalian		
Frog	93	43
Chick	—	130
C. Mammalian		
Rats—postimplantation	97	38

SOURCE: Adapted from Daston and D'Amato (*18*).

A second approach, used by Fabro and Brown (27) and Johnson (28), has been the determination of the ratio between the concentration of a compound that causes toxicity in an adult to one that causes developmental toxicity in an embryo or fetus. Fabro and Brown (27) advocated using a relative teratogenic index (RTI), which they calculated by dividing the adult dose LD_1 (the concentration that kills 1% of adults) by the teratogenic dose TD_5 (the concentration that causes teratogenesis in 5% of embryos). Johnson (28) proposed an A/D ratio, which he defined as the lowest observed adverse effect level (LOAEL) of the adult toxic dose (A) divided by the LOAEL for developmental toxicity (D). Recently, this was modified by dividing the adult no adverse effect level (NOAEL) by the developmental NOAEL (29). The RTI and A/D ratio are fundamentally similar, but the RTI is a statistical value calculated from a dose–response curve, whereas the A/D ratio is a ratio of LOAELs in which no consideration is given to the slope of the dose–response curve. Both ratios are based on the concept that most teratogens are specifically toxic to the embryo and much less toxic to adults. Johnson stated that a critical A/D ratio of 3 or higher is indicative of a reproductive hazard (28). Teratogens, therefore, would have a high RTI or A/D ratio. In practice, as far as human teratology is concerned, this may not be true, because certain chemotherapeutic agents, anticonvulsants, and ethanol are only teratogenic at or near levels that cause maternal toxicity.

Tissue Culture. Cell lines have been established from mouse ovarian tumor (MOT) cells (30), human embryonic palatal mesenchyme cells (31), and human neuroblastoma cells (32). Cell attachment, growth inhibition, and inhibition or stimulation of differentiation are the respective parameters measured after exposure to nontoxic doses of the test chemical.

The rationale of the MOT assay is based on the generic property of embryonal cells to adhere to surfaces or lectins. The assay is performed by adding the test chemical to a suspension of MOT cells pretreated with radiolabeled thymidine. A plastic sheet coated with lectin is placed in the culture bottle so that the MOT cells can adhere to it. Inhibition of adherence is measured by determining the radioactive counts associated with the plastic sheet and the medium and by comparing these values with the control. In a study of more than 100 teratogens and nonteratogens, a 79% accuracy compared with in vivo results was obtained.

The assay in which human embryonic palatal mesenchyme cells are used is based on the premise that teratogens will inhibit the growth of these rapidly dividing cells. Growth inhibition is measured by cell counting. A critical concentration of 1 mM was chosen for the uppermost concentration in the assay. An evaluation of known teratogens and nonteratogens showed that 66% of the teratogens were correctly identified as positives and that 60% of the nonteratogens were identified as true negatives (33).

The third assay uses human neuroblastoma cells derived from childhood malignant tumors that in culture can differentiate into neurites. Of 42 teratogens tested, 76% stimulated differentiation and 7% inhibited differentiation, and of the nonteratogens tested, 74% did not alter differentiation (34).

Organ Culture. *Mouse Embryo Limb Bud Cell Culture Assay.* In this assay, 11- to 12-day-old mice embryos are removed from the dams and the limb bud dissected away from the embryo under a stereomicroscope (35). The buds are incubated for several days in a nutrient medium in the presence or absence of the test chemical. The cultured limbs can then be scored qualitatively for the presence or absence of malformations and quantitatively for the amount of cartilage formed in the limb by staining with toluidine blue. A quantitative approach to the measurement of the formation of cartilage is the determination of the uptake of ^{3}H-proline or ^{35}S-sulfate into chondroitin sulfate, the major component of cartilage. A modification of this method, developed by Guntakatta et al. (36), depends on the disassociation of the limb bud into individual cells, which are then exposed to ^{3}H-thymidine and ^{35}S-sulfate in culture. An overall accuracy of 89% was reported by Guntakatta et al. (36) after the testing of both teratogens and nonteratogens that do not require metabolic activation. An advantage of this assay is the fact that the limb buds can be cultured for 9 days, during which time cartilagenous bone structures develop. This permits the evaluation of a number of end points, including chondrogenesis and collagen biosynthesis (35). Metabolizing enzymes added either as cell-free extracts or intact cells have successfully been used in the assay.

Whole Embryo Systems. *Invertebrate Embryos.* Hydra, a test system, which was developed by Johnson and co-workers (28, 37, 38), utilizes the freshwater coelenterate *H. attenuata* and is based on the reaggregation and differentiation of the coelenterate. Adult hydra polyps are exposed to log concentrations of the test substance (1×10^{-3} to 1×10^3 μL/mL) and the minimum concentration required to produce a toxic response is determined. The dose of a test compound that results in toxicity in the adult (A) is then compared with that which causes developmental toxicity (D). The embryo is prepared by disassociating 700–1000 hydra into their component cells to form an "artificial embryo". These preparations consist of two cell types, the major component of which consists of groups of cells that give rise to a new population of adult polyps in about 90 h. The second group consists of a small number of undifferentiated interstitial cells capable of rapid proliferation and subsequent differentiation into a typical adult hydra. The concentration that prevents the formation of the adults from the undifferentiated cells is the dose that causes developmental toxicity. An A/D ratio can then be calculated. If this ratio is greater than 3, it has been suggested that the compound is a reproductive toxicant. More than 30 compounds that have been tested in this assay had in vitro A/D ratios that compared well with in vivo A/D ratios (38, 39). Thus, this assay seems to be extremely promising in predicting the reproductive toxicity of certain compounds.

Planaria, a system devised by Best and Morita (40) using the freshwater planarian *Dugesia dortocephala*, has been used to assess the effects of toxicants on reproduction and teratology. Planaria exhibit bilateral symmetry, are hermaphroditic, can reproduce sexually or asexually, and have a true brain indicative of complex behavior. Planaria reproduce by fission in the transverse plane with regeneration of the anterior or posterior section. Regeneration involves the migration of undifferentiated stem cells or neoblasts. Because this process resembles embryogenesis, it is thought to be sensitive to teratogenic agents. Two experimental approaches have been developed: (1) Surgical fragments of planaria can be exposed to the test material during regeneration to assess the effect of the compound on the rapid multiplication of cells, or (2) intact planaria can be exposed to the test material and examined for morphologic or behavioral abnormalities. This assay has been shown to be less sensitive to three known teratogens than the hydra assay (41).

Lower Vertebrate Embryos. Fish, (Japanese medaka, rainbow trout, fathead minnow, and zebra fish), have been used to test for adverse reproductive outcomes, but only a few compounds have been evaluated in fish. The medaka seems to be the most promising species, because it is small, oviparous (producing eggs that hatch outside the body), and easy to culture in the laboratory; much of its biology is known; and it has a clear chorion (outer embryonic membrane) so that adverse changes can easily be seen at various stages of development. A number of end points can be monitored, including death, specific malformations, delayed hatching, and functional ones, such as impaired swimming movements. Exposure can be done directly into the water (42) or by microinjection of the embryo at various stages of development (43). The most extensive studies have been undertaken by Birge et al. (42), who found rainbow trout to be the most sensitive to teratogenic agents, followed by goldfish, sunfish, and largemouth bass. Irrespective of whether exposure was to inorganic or organic compounds, the majority of the malformations involved the skeletal system.

In frogs, the frog embryo teratogenesis assay (known as FETAX), uses the South African clawed toad, *Xenopus laevis*. Fertilized eggs are exposed to the toxicant in question for the first 96 h of development, and the medium is replaced every 24 h. After 4 days, the surviving embryos are fixed in formalin, malformations noted, and the developmental stage determined (44). The number of dead embryos are recorded daily. A 96-h LC_{50} (mortality) and EC_{50} (concentration inducing terata in 50% of the surviving embryos) are determined. The teratogenic index (LC_{50}/EC_{50}) is calculated. Both methotrexate and 5-azacytidine, which are highly teratogenic in FETAX, have also been shown to be teratogenic in mammalian and in human studies. Aspartame and amaranth, which are both weak teratogens or negative in FETAX, have not been shown to be teratogenic in humans (45). Eye defects and anencephaly (lack of cerebral hemispheres) have been observed in both mammalian assays and FETAX (46, 47) following exposure to methotrexate. In humans, absence of frontal bones and premature closing of the brain sutures have been documented in children whose mothers were exposed to methotrexate (48). It has been shown that cytochrome P-450 enzymes can be added to the FETAX assay to metabolize teratogens such as diphenylhydantoin to active intermediates. Craniofacial defects have been found in children born to mothers treated with hydantoin for epilepsy during pregnancy (49).

Mammalian Embryos. In the preimplantation assay early mammalian embryos are collected through the blastocyst stage and cultured in in vitro in serum-amended media (50, 51). Exposure of the embryo can be undertaken during culturing or by exposure of the mother before the embryo is flushed from the reproductive tract. Morphological alterations in development to the blastula stage, hatching from the zona pellucida, attachment to a culture dish, and outgrowth of the three types of cells, are the end points measured in the assay. To date, only a few compounds (cyclophosphamide, cadmium chloride, irradiation, methylmercury, mercuric chloride, potassium dichromate, nitroquinoline N-oxide, mitomycin C, bleomycin, diethylstilbestrol, and ochratoxin A) have been evaluated for reproductive toxicity using this assay (52). Cytological end points, such as sister chromatid exchanges and chromosomal aberrations, have been detected in the embryos that have shown adverse effects. Extrapolation of these findings to humans is difficult, because under normal conditions, only 10% of embryos in the mouse are lost during the first few days after fertilization, whereas in humans, this may be as high as 75% (53).

The postimplantation assay, which was originally developed by New (54), involves the removal of 9- or 10-day-old (equivalent to 4–12 somites) mice or rat embryos from the uterus followed by in vitro culturing in rat and human serum for a period ranging from 1 to 48 h (55–57). During incubation, test chemicals are added to the medium. After exposure, embryos are examined for changes in heartbeat, yolk sac circulation, crown rump length, somite numbers, closed or open neural tubes, otic and optic vesicles, and development of limbs. A morphological scoring system for teratogenic changes in rats has been developed by Brown and Fabro (58). In addition to morphological characteristics, various biochemical parameters, including protein content, DNA content, uptake of ^{3}H-thymidine and ^{14}C-amino acids into DNA and proteins, respectively, can be measured.

A study by Schmid (59), in which actinomycin D, azathioprine, colchicine, coumarin, doxorubicin, hycanthone, ketoconazole, methotrexate, and trypan blue were used, demonstrated that the number of somites after exposure was as reliable an index of teratogenicity as any other morphological change. However, the author also determined that each compound elicited a specific response resulting in a defect at a particular embryonic sites. The concentrations that elicited a teratogenic response ranged from 3 ng/mL (actinomycin D) to 150 µg/mL (trypan blue). Results of this study (59) were similar to those obtained in in vivo studies with actinomycin D (60), hycanthone (61), and methotrexate (62). Although coumarin was shown to cause central nervous system malformations in this assay (89) and has been implicated in congenital defects of the central nervous system in humans (63, 64), it was negative in in vivo studies in rabbits (65) and mice (66).

Advantages and Disadvantages of In Vitro Assays. In vitro assays have many advantages as well as a number of inherent problems. Advantages are the relatively low costs, the simplicity and the rapidity with which results are obtained, the many permutations that can be used for the scientific protocol, the precise control over exposure to a specific developmental stage of the embryo, the composition of the culture medium, the concentration of the test compound, and the duration of exposure. The major disadvantages include the isolation of the embryo from the mother, thus eliminating the modulation of the maternal metabolic system and the difficulty in assessing the long-term consequences of brief embryonic exposures used in these assays. Additional problems include the lack of standardization of the methods and quantitation of normal and abnormal embryonic development. The influence of chronic exposure cannot be analyzed by any of these methodologies, and the extrapolation of the results found in laboratory animals to that of humans is difficult. However, in vitro assays do have potential value for the future, as a good correlation between in vitro and in vivo studies for a few compounds has been documented.

In Vivo Mammalian Assays. *Mammalian Three-Generation Studies.* In these reproductive toxicity tests, three generations of animals are exposed to the toxic agent in question at various dose levels. The test substance is administered to both parents throughout the study; the highest dose level is toxic to the test animal and the lowest is innocuous (67). Following exposure of the animals during gestation, various developmental parameters are measured in the pups. Offspring of the mating are continuously exposed to the compound, weaned, allowed to mature, and mated among themselves. This procedure is followed for three succeeding generations. The advantage of this technique is the determination of the effect during in utero exposure and subsequent reproductive performance. A complication that does arise is the selection of the animals for mating for the succeeding generations. Fetal survival, litter size, and weight of the pups may affect the selection process. At the end of each reproductive cycle, all the pups are examined for physical abnormalities. The number of viable, stillborn, and dead pups in each litter is recorded, and the numbers of survivors on days 1, 4, 7, 14, and 21 are recorded. Body weight of the pups is recorded at weaning. Two indices, the gestation index and the viability index (VI), can be calculated. The gestation index, which is a measure of the percentage of pregnancies resulting in live litters, is not a meaningful measure, because it does not take into account stillborn pups unless the whole litter succumbs. The VI refers to the percentage of pups that survive for a specific period. This index is extremely important, because early survival of the pups may be dependent on the excretion of a toxic compound into the milk of the lactating dam and subsequent ingestion by the pups.

In these three-generation studies, a number of parameters cannot be measured. For example, the actual litter size is seldom known, because the mothers are not sacrificed after parturition. No information is collected on sperm motility or viability. Groups of untreated control animals must be included in these studies so that comparisons with the exposed group can be made. The reproductive organs of only pups not being used for breeding can be evaluated by weighing or by examination for histological changes. In most studies, only the male reproductive organ, that is, the testicle, is studied, but in some situations, the ovary may be examined.

Mammalian Single-Generation Studies. In single-generation reproductive studies, both male and female animals receive the compound for 60 days prior to conception and then through gestation. The effect on libido, estrus cycle, reproductive capability, toxicity of the compound to both mother and fetus, postnatal pup development, and adequacy of lactation in the mother can be evaluated. In these conventional reproduction studies, the animals are mated once or twice following exposure to the test chemical. A new protocol, known as the Fertility Assessment by Continuous Breeding, which has been developed by the National Toxicological Program (NTP), allows multiple matings for each pair of animals during the 14 weeks of cohabitation and exposure to the toxicant. In these studies, females generally deliver four to five litters during the exposure period; thus, a meaningful fertility index can be calculated (68).

Another approach, one that is used more frequently, is the administration of the toxicant to the mother following mating. Several dose levels of the chemical are administered to different groups of females. The highest dose is based on the maximum tolerated dose (MTD) and frequently results in maternal toxicity, while the lower doses are set at specific fractions of the MTD. To prevent cannibalism of the pups, the fetuses are usually delivered by caeserean section prior to their expected date of birth. The newborn pups are examined for external and internal abnormalities. A major problem associated with this approach is that teratogenic changes that will occur later in life are not detected. A modified experimental protocol that has been used by many investigators is exposure of the rodents on days 1 through 19 of gestation to the toxicant followed by sacrifice of the dams and removal of the fetuses for morphological and sometimes histological evaluation. Frequently, the route of exposure is similar to that which humans would encounter, but the outcomes are different.

Effects of Chemicals on Reproductive Outcomes in Laboratory Animals.
Single-generation assays have found many common solvents to be teratogenic or embryotoxic in rodents. These include benzene (69), xylene (70), cyclohexanone (71), propylene glycol (72), alkane sulfones (73), glycol ethers (74), acetamides, and formamides (75).

In general, insecticides appear to have low embryotoxic potential in mammals, but a number of teratogenic effects have been recorded with organochlorine insecticides, including aldrin, dieldrin, and endrin in the mouse and hamster (76). A few carbamates and organophosphates, such as carbaryl in the guinea pig (77) and dog (78), demeton (79), and parathion (80) in the mouse, and dichlorvos in the rabbit (81), are known to be teratogenic. Evidence that the fungicides mancozeb, dinocap, and nitrofen are potent teratogens has been presented (82). DDE, a metabolite of 1,1,1-trichloro-2,2-bis(*p*-chlorophenyl)ethane (DDT) is known to have detrimental effects on reproduction in birds by interfering with the hormonal activity of estrogen, which results in decreased deposition of calcium in the eggshell. The hormone methyltestosterone has been shown to be teratogenic in dogs, resulting in pseudohermaphroditism in female offspring (83). A study by Nelson et al. (84) that assessed the developmental toxicity of 13 alcohols showed that teratogenic effects were not elicited by low-molecular-weight alcohols (methanol and ethanol), as had been originally hypothesized. Also, the concentrations that elicited effects exceeded 5000 ppm and caused maternal toxicity.

Epidemiological Studies in Humans

Epidemiological studies for reproductive loss are usually undertaken using retrospective approaches, including personal interviews and medical records. If only personal interview data are used, many confounding factors can interfere with an accurate assessment of a spontaneous abortion. Patients are frequently confused about delayed menstruation and cannot recall events. The confirmation of a miscarriage through medical documentation is thus necessary to obtain relevant data (67, 85). Selection bias of exposed and nonexposed women must also be avoided in these studies. A number of factors not involving exposure have also been associated with spontaneous abortion. These include maternal age, previous spontaneous abortions, cervical incompetence, maternal fever (78, 85, 86), diet, health status, and weight (87).

In an epidemiological study, evaluation of timing and duration of exposure is crucial. Exposure of the father for 4 months prior to conception must be monitored, because this is the interval required for a complete cycle of spermatogenesis. Possible fetal exposure during the first trimester is the most important, although exposure during the second and third trimesters can result in adverse effects. Quantification of exposure is frequently the most difficult parameter to measure. An innovative approach that has been applied in assessing exposure has been biological markers, which are noninvasive measures of dose that can be related to potential adverse reproductive outcomes (88). Urinary mutagens, thioethers, or D-glucaric are frequently measured biomarkers. However, use of these biomarkers requires information about their normal values, inter- and intravariability among individuals, and possible confounding effects.

To assess the effect of exposure on reproductive outcomes, two sequential epidemiological techniques, descriptive and analytical have been used (89, 90). In descriptive studies, information regarding the distribution and frequency of an outcome is obtained. The impetus for these studies is usually based on an increased number of case reports on spontaneous abortion or teratology in a particular geographical area. The second stage is an analytical study designed to test a hypothesis or to generate a new hypothesis about an association between exposure and reproductive outcomes. Either case-control or cohort studies are used. In the case-control study, a retrospective assessment of the exposure factors of the cases and controls is performed. In a cohort study, both individuals exposed and not exposed to a particular factor are followed over time; either retrospective or prospective techniques are used to observe the outcome of interest.

Chemicals Causing Spontaneous Abortion. A number of chemicals have been implicated as being abortifacient in humans, although definitive epidemiological data on many compounds are still lacking (91). However, there is some evidence that arsenic, lead, and anesthetic gases can cause spontaneous abortion, particularly during the early weeks of pregnancy. Extensive epidemiological studies in Sweden by Norstrom and co-workers (92–94), have reported an increased prevalence of spontaneous abortion in women living close to or working in a smelter that discharges arsenic and lead. Decreased birth weight of offspring not associated with cigarette smoking was shown (94).

Health providers seem to be at greater risk of adverse reproductive outcomes than many other occupationally exposed groups. For example, epidemiological studies have shown that exposure of operating room nurses and female anesthesiologists (95, 96) to anesthetic gases resulted in an increased rate of spontaneous abortion, compared with unexposed control women. Also, exposure of nurses during the first trimester to antineoplastic drugs, including cyclophosphamide, doxorubicin, and vincristine, has resulted in an increased incidence of fetal loss (97). The independent effect of each drug could not be proved, because more than one drug was handled by each nurse during the exposure period.

In Finland, an increase in spontaneous abortion was also observed in women involved in the sterilization of hospital instruments using ethylene oxide (98). Exposure of personnel ranged from 5 to 10 ppm, there were occasional acute concentrations of 250 ppm. Neither teratogenic effects nor decreased birth weights were observed. Although ethylene oxide ranks 26th in the volume of organic chemicals produced in the United States, less than 1% is used in hospitals (5).

A second cohort that may have a higher risk of spontaneous abortions are workers exposed to pesticides. A significant increase in abortions among occupationally exposed couples in India has been reported (99–101). These workers were exposed to nine pesticides, including DDT, dieldrin, dichlorvos, parathion, metasystox, and copper sulfate. An increase in chromosomal aberrations in lymphocyte was also documented in these sprayers (102).

In Sweden, a statistically higher rate of miscarriage was found among women working at a petrochemical plant, but no differences were observed in women living close to the plant (103). A similar nonsignificant increase in reproductive outcomes was noted in a group of women living close to a phenoxy herbicide plant (104).

Chemicals Causing Teratology. Teratogenic effects are usually manifest at birth, but with some compounds an adverse response is noted only during the teens or early twenties. For example, prenatal exposure of males to diethylstilbestrol, a drug used to prevent miscarriage, resulted in abnormalities in the male reproductive tract, manifested as epididymal cysts, seminal vesicle enlargement, prostatic inflammation, undescended testes, and the production of abnormal sperm (105). By interfering with the male hormones necessary for development and differentiation of the male reproductive tract, prenatal diethylstilbestrol exposure was responsible for morphological abnormalities expressed in later life. In the case of the female fetus exposed in utero to diethylstilbestrol, adenocarcinoma, a malignant cancer of the reproductive tract, has been diagnosed in a number of young women.

Mercury is the metal for which the greatest documentation of teratology has been recorded. During the period of 1954–1960, the population of Minamata Bay in Japan was exposed to methylmercury via the food chain. A chlor-alkali plant, in which mercury was used as a catalyst, discharged its wastes into the bay. Subsequent methylation of the mercury by bacteria and uptake of the lipophilic methylmercury by fish resulted in bioconcentration of the compound into fat and muscle. Because fish was the staple source of protein for the inhabitants of the area, a number of people were exposed to the compound. Infants exposed in utero via the consumption of contaminated fish by their mothers demonstrated varying degrees of neurological symptoms resembling cerebral palsy (106, 107). Similar effects of mercury on development has also been documented in the United States (108) and Iraq (109) following accidental inclusion of mercurous fungicides into grains.

Other metals for which epidemiological evidence exists regarding their teratogenic effects are aluminum and lithium. In a study in South Wales, a significant positive correlation between central nervous system malformations and aluminum was noted (*110*). Lithium in the form of lithium carbonate, which is used therapeutically to treat psychiatric disorders, has been associated with teratogenicity in infants born to mothers exposed during the first trimester of pregnancy (*111, 112*).

Accidental exposure of more than 2000 people to cooking oil contaminated with polychlorinated biphenyls (PCB) and dibenzofurans during a 6-month period in Taiwan resulted in teratogenicity. Malformed children were still being born to mothers exposed more than 6 years previously to the cooking oil (*113*). This long-lasting phenomenon may be related to the bioaccumulation in fatty tissue of the ingested PCB, which in humans has an estimated half-life of 7 years. Abnormalities of the gingiva, skin, nails, teeth, and lungs were observed in a statistically larger percentage of offspring of exposed mothers. Neurologically, the exposed infants did not differ from the unexposed controls, but a delay in the performance of certain tasks requiring motor coordination was observed. Also, the exposed children scored lower on three IQ tests. An earlier study in Japan with a smaller cohort of children whose mothers had been exposed to thermally degraded cooking oil showed similar abnormalities (*114*). At 7 years of age, 70% of these Japanese children were apathetic and had IQs in the 70s.

Conflicting reports on the teratogenic potential of 2,4,5-trichlorophenoxyacetic acid (2,4,5-T) and its major contaminant 2,3,7,8-tetrachlorodibenzodioxin has appeared in the literature (*5*). An extensive study, by the U.S. Department of Defense on 0.5 million exposed Vietnamese, reported no increased incidence of either miscarriages or teratogenic outcomes (*115*), whereas an American Association for the Advancement of Science (AAAS) report showed an increase in cleft palate and spina bifida in children in Vietnam from 1964 to 1968 (*116*). A study by LaPorte (*117*) substantiated the results of the AAAS, whereas a study by Kundstadter (*118*) found no increased incidence from the retrospective examination of hospital records.

Life-style factors, such as smoking and consumption of ethanol, have been clearly implicated as teratogenic agents. Smoking during pregnancy has been associated with reduced birth weight of infants (*119*) and an increased risk of spontaneous abortion (*85,120*). This effect, which has been coupled to growth retardation, has been found to be more pronounced in blacks than whites (*121*). A prevalent syndrome, known as fetal alcohol syndrome, has been diagnosed in infants whose mothers consumed more than two drinks per day during pregnancy (*122*). These children have varying craniofacial malformations including midfacial and maxillary hypoplasia (incomplete development), microencephaly (small brain), abnormal joints, and heart abnormalities. They also have severe growth deficiencies and intellectual handicaps (*123*).

Drugs that are known human teratogens include thalidomide; androgens; aminopterin (*2*); coumarin derivatives; trimethadione, an anticonvulsant; and isotretinoin, used in acne treatment (124). Isotretinoin contains an aromatic retinoid, etretinate, used to treat skin diseases. Malformations in newborns manifested as craniofacial and limb defects were reported to be associated with intake of etretinate by pregnant mothers (125). Three aborted fetuses examined by these investigators were found to have severe central nervous system anomalies.

Chemicals Affecting the Normal Production of Spermatozoa. Male-mediated adverse reproductive outcomes result either from direct chemical effects on the testes and germ cells or through interference with the steroidogenic function of the testes. Dibromochloropropane (DBCP), a potent alkylating agent that was developed as a pesticide, is known to inhibit (azoospermia) or decrease (oligospermia) the development of sperm. Reduced sperm counts, testicular atrophy, and decreased fertility were documented in male employees in a plant manufacturing DBCP (126,127). Similar reproductive effects were noted in occupationally exposed workers applying the nematocide (128). Subchronic exposure of laboratory animals to DBCP produced similar changes to those observed in humans (129).

A second compound that alters sperm production is lead, which produces malformed sperm (teratospermia) in the ejaculate (130). However, no relationship between the abnormally shaped sperm and adverse reproductive outcomes have been established (131). Deleterious changes in lymphocytes and sperm have been documented among male workers exposed to arsenic and lead at a smelter, there were respective increases in chromosomal aberrations in cultured lymphocytes and impaired morphology and motility of sperm (132). Because these workers were concomitantly exposed to cadmium and other smelter fumes, the effects may not be due to a particular metal alone but rather due to a synergistic interaction between more than one toxin simultaneously.

Extrapolation from Animal Studies to Humans. In extrapolating from animal studies to humans, a number of factors must be taken into consideration: The appropriate laboratory animal must be utilized, the number of animals on each treatment must allow for statistical evaluation, and the dose levels should be relevant to human exposure.

Far more animal than human teratogens have been identified because of the way in which we are able to test animals. Experimental animals have a short reproductive cycle and multiple offspring. Therefore, a large number of fetuses can be examined in a short time and at relatively low cost. An equivalent human epidemiological study would cost millions of dollars and require a number of years. The dosage administered to animals can be adjusted so that it is many times greater than would be expected in a normal human exposure situation. The experiment may also be designed to correlate exposure of the embryo or fetus with its most sensitive time in development (1). All of these factors improve the possibility of identifying a chemical compound as a teratogen in animals.

Animal studies are, however, not without fault. The variation in genotype among different individuals influences their response to a teratogen and hence their susceptibility to a particular toxicant. Even greater variations in genotype exist among different species. Therefore, a chemical proved to be nonteratogenic in a certain laboratory species may well be teratogenic in a different laboratory species or even in humans. A classic example is that of thalidomide. The laboratory animals most widely used in teratogenicity testing (mouse, rat, and rabbit) are 5–500 times less sensitive than humans to this sedative (5), which is one of the most potent human teratogens known.

From a risk perspective, however, animal studies have generally been good predictors of teratogenicity. Studies by Schardein (133), Fabro (134), and Jelovsek et al. (135) have indicated that human teratogens yield positive results in animal and in vitro studies more often than do nonteratogens. It has been estimated that 75% of human developmental toxicants can be predicted on the basis of animal studies only (136), but unfortunately false-negative results are also high (25%).

Other factors for consideration are the combinations and permutations of chemicals after they enter the environment. Even if a chemical is found to be nontoxic in animal studies, the safety of the chemical can not be ensured. The possible potentiation of a chemical's effect by its interaction with other chemical agents or other toxic factors once it enters the environment cannot be dismissed.

Conclusion

It is evident that many possible adverse health effects may result from exposure to the numerous chemicals in the environment. Exposure of both male and female organisms to chemicals during the reproductive cycle can have deleterious effects on the development of the fetus. These effects are manifested as fetal death, malformation, retarded growth, and organ dysfunction. Our knowledge of the adverse health effects that are produced as the result of a complex network of factors is limited. This is the key point: Much is unknown about the chemicals to which humans are exposed and how they interact with biological systems. There is much progress to be made in defining the incidence of such effects, the magnitude of dose in specific exposure situations, and the possibility of interactions among causes. To aid in assessing whether a compound is a teratogen, a number of short-term in vitro assays have been developed and are presently being validated. The progress made so far has indicated the importance of limiting exposure to chemicals to the most practical and feasible extent. Future research efforts should be directed toward establishing the mechanisms of toxicological effects and developing more accurate means of identifying hazardous compounds.

References

1. Shepard, T. H. *Ann. N. Y. Acad. Sci.* **1986**, *477*, 105–115.
2. Wilson, J. G. In *Handbook of Teratology:* Wilson, J. G.; Fraser, F. C., Eds.; Plenum: New York, 1977; Vol. 1, pp 357–385.
3. Byrn, F. W.; Gibson, M. *Clin. Obstet. Gynecol.* **1896**, *29*, 925–940.
4. Stellman, J. M. *Ann. N. Y. Acad. Sci*, **1986**, *477*, 116–121.
5. Schardein, J. L.; Keller, K. A. *CRC Crit. Rev. Toxicol.* **1989**, *19*, 251–339.
6. Butcher, R. L.; Page, R. D. *Environ. Health Perspect.* **1981**, *38*, 35–37.
7. Pernoll, M. L. *Clin. Obstet. Gynecol.* **1986**, *29*, 953–958.
8. *Dorland's Illustrated Medical Dictionary*, 25th ed.; Friel, J. P., Ed.; WB Saunders: Philadelphia, PA, 1974.
9. Wilson, J. G. In *Handbook of Teratology;* Wilson, J. G.; Fraser, F. C., Eds.; Plenum: New York, 1977; Vol. 1, pp 49–74.
10. Langman, J. *Medical Embryology*, 2nd ed.; Williams & Wilkins: Baltimore, MD, 1969.
11. Taskinen, H.; Lindbohm, M. L.; Hemminki, K. *Br. J. Ind. Med.* **1986**, *43*, 199–205.
12. Johnston, S. D.; Raksil, S. *Vet. Clin. North Am.: Small Anim. Pract.* **1987**, *17*, 535–554.
13. Carr, D. H. In *Handbook of Teratology.* Wilson J. G.; Fraser, F. C., Eds.; Plenum: New York, 1977; Vol. 3, pp 200–213.
14. Sweeney, A. M.; Meger, M. R.; Aarons, J. H.; Mills, J. L.; La Porte, R. E. *Am. J. Epidemiol.* **1988**, *127*, 843–850.
15. Beck, F. In *Developmental Toxicology;* Kimmel, C. A.; Buelke-Sano, J., Eds.; Raven: New York, 1981; pp 35–54.
16. Wilson, J. G. *Fed. Proc.* **1977**, *36*, 1698–1703.
17. Faustman, E. M. *Mutat. Res.* **1988**, *205*, 355–384.
18. Daston, G. P.; D'Amato, R. A. *Toxicol. Ind. Health* **1989**, *5*, 555–585.
19. Fantel, A. G.; Greenway, J. C.; Juchau, M. R.; Shephard, T. H. *Life Sci.* **1979**, *25*, 67–72.
20. Hales, B. F. *Teratology* **1981**, *24*, 1–11.
21. Steele, V. E.; Morrisy, R. E.; Elmore, E. L.; Gurganus-Rocha, D.; Wilkinson, B. P.; Curren, R. D.; Schmetter, B. J.; Lovie, A. T.; Lamb, J. C., IV; Yang, L. L. *Fund. Appl. Toxicol.* **1988**, *11*, 673–684.
22. National Toxicology Program. *Final Report NTP 186*, 86–372.
23. Oglesby, L. A.; Ebron-McCoy, M. T.; Logsdon, T. R.; Copeland, F.; Beyer, P. E.; Kavlock, R. J. *Teratology* **1992**, *45*, 11–33.
24. Mayura, K.; Smith, E. E.; Clement, B. A.; Phillips, T. D. *Toxicol. Appl. Pharmacol.* **1991**, *108*, 253–266.
25. Kistler, A. *Arch. Toxicol.* **1987**, *60*, 403–414.
26. Rawlings, S. J.; Shuker, D. E. G.; Webb, M.; Brown, N. A. *Toxicol. Lett.* **1985**, *28*, 49–58.
27. Fabro, S.; Brown, N. A.; *Teratogenesis, Carcinog. Mutagen.* **1982**, *2*, 61–76.
28. Johnson, E. M. *Ann. Rev. Pharmacol. Toxicol.* **1981**, *21*, 417–429.
29. Johnson, E. M. *Teratology*, **1987**, *35*, 405–427.
30. Braun, A. G.; Emerson, D. J.; Nicison, B. B. *Nature (London)* **1979**, *282*, 507–509.
31. Pratt, R. M.; Grove, R. I.; Willis, W. D. *Teratogenesis Carcinog. Mutagen.* **1982**, *2*, 313–318.
32. Mummery, C. L.; Van Den Brink, C. E.; Van Der Saag, P. T.; De Laat, S. W. *Toxicol. Lett.* **1983**, *18*, 201–209.
33. Braun, A. G.; Buckner, C. A.; Emerson, D. J.; Nichison, B. B. *Proc. Natl. Acad. Sci. U.S.A.* **1982**, *79*, 271–279.

34. Mummery, C. L.; Van Den Brink, C. E.; Van Der Saag, P. T.; De Laat, S. W. *Teratology* **1984**, *29*, 271–279.
35. Kochar, D. M. *Teratogenesis Carcinog. Mutagen.* **1982**, *2*, 303–312.
36. Guntakatta, M.; Matthews, E. J.; Rundell, J. O. *Teratogenesis Carcinog. Mutagen.* **1984**, *4*, 349–364.
37. Johnson, E. M. *J. Environ. Pathol. Toxicol.* **1980**, *4*, 153–156.
38. Johnson, E. M.; Gorman, R. M.; Gabel, B. E. G.; George, M. E. *Teratogenesis Carcinog. Mutagen.* **1982**, *2*, 263–276.
39. Johnson, E. M.; Gabel, B. E. G.; Larson, J. *Environ. Health Perspect.* **1984**, *57*, 135–139.
40. Best, J. B.; Morita, M. *Teratogenesis Carcinog. Mutagen.* **1982**, *3*, 277–291.
41. Sabourin, T. D.; Faulk, R. T.; Yos, L. B. *J. Appl. Toxicol.* **1985**, *5*, 227–233.
42. Birge, W. T.; Black, J. A.; Westerman, A. G.; Ramey, B. A. *Fund. Appl. Toxicol.* **1983**, *3*, 237–242.
43. Medcalfe, C. D.; Sonstegard, R. A. *J. Natl. Cancer Inst.* **1985**, *75*, 1091–1097.
44. Fort, D. J.; Bantle, J. A. *Fund. Appl. Toxicol.* **1990**, *14*, 720–733.
45. Holson, J. F.; Gaines, T. B.; Schumscher, H. J.; Cramer, M. F. *Toxicol. Appl. Pharmacol.* **1975**, *33*, 122.
46. Kronick, J. B.; Whelan, D. T.; McCallion, D. J. *Teratology* **1987**, *36*, 245.
47. Bantle, J. A.; Fort, D. J.; Rayburn, J. R.; De Young, D. J; Bush, S. J. *Drug Chem. Toxicol.* **1990**, *13*, 267–282.
48. Milunsky, A.; Graef, J. W.; Gaynor, M. F. *J. Pediatr.* **1968**, *72*, 790.
49. Hanson, J. W.; Smith, D. W. *J. Pediatr.* **1985**, *87*, 285–290.
50. Spielmann, H.; Kruger, C.; Vogel, R. *Concepts Toxicol.* **1985**, *3*, 22–28.
51. Spielmann, H.; Kruger, C.; Tenschart, B.; Vogel, R. *Drug Res.* **1986**, *36*, 219–223.
52. Spielmann, H.; Vogel, R. *Crit. Rev. Toxicol.* **1989**, *20*, 51–64.
53. Kline, J.; Stein, Z. In *Reproductive Toxicology;* Dixon, R. L., Ed.; Raven: New York, 1985; p 251.
54. New, D. A. T. *Biol. Rev.* **1978**, *53*, 81–122.
55. Fantel, A. G. *Teratogenesis Carcinog. Mutagen.* **1982**, *2*, 231–242.
56. Kochar, D. M. *Teratogenesis Carcinog. Mutagen.* **1980**, *1*, 63–74.
57. Van Maele-Fabry, G.; Picard, J. J. *Teratology* **1987**, *36*, 95–106.
58. Brown, N. A.; Fabro, S. *Teratology* **1981**, *24*, 65–78.
59. Schmid, B. P. In *Concepts in Toxicology;* Homburger, F.; Goldberg, A. M., Eds.; S. Karger: Basel, Switzerland, 1985; Vol. 3, pp 46–57.
60. Tuchmann-Duplessis, H.; Mercier-Parot, L. In *Ciba Foundation Symposium on Congenital Malformation;* Wolstenholme, C., Ed.; Little, Brown: Boston, MA, 1958; pp 115–128.
61. Moore, J. A. *Nature (London)* **1972**, *239*, 107–109.
62. Wilson, J. G.; Scott, W. J.; Ritter, R.; Fradkin, R. *Teratology,* **1979**, *19*, 71–80.
63. Warkany, J. *Teratology* **1976**, *14*, 205–209.
64. Kleinbrecht *Dtsch. Med. Wochenschr,* **1982**, *107*, 1929–1931.
65. Grote, W.; Weinmann, I. *Arzneimittelforschung* **1973**, *23*, 1319–1320.
66. Kronick, J.; Phelps, N. E.; McAllion, D. J.; Hirsh, J. *Am. J. Obstet. Gynecol.* **1974**, *118*, 819–823.
67. *Test Guidelines: Health Effects;* Annual Report; Office of Pesticides and Toxic Substance, Environmental Protection Agency: Washington, DC, August 1992; EPA 560/6–82–000, p 432.
68. Lamb, J. C. *J. Am. Coll. Toxicol.* **1985**, *4*, 163–171.
69. Watanabe, G.; Yoshida, S.; Hirose, K. *Proceedings of the Congenital Abnormalities Research Association;* Tokyo, Japan, 1968; p 45.
70. Kuchera, J. J. *Pediatrics* **1968**, *72*, 857–859.
71. Weller, E. M.; Griggs, J. H. *Teratology* **1973**, *7*, 1–30.

72. Gebhardt, D. O. E. *Teratology* **1968**, *1*, 153–162.
73. Hemsworth, B. N. *J. Reprod. Fertil.* **1968**, *17*, 325–334.
74. Schuler, R. L.; Hardin, B. D.; Niemeyer, R. W.; Booth, G.; Hazeldon, K.; Piccirillo, V.; Smith, K. *Environ. Health Perspect.* **1984**, *57*, 141–146.
75. Thiersch, J. B. *J. Reprod. Fertil.* **1962**, *4*, 219–220.
76. Ottolenghi, A. D.; Haseman, J. K.; Suggs, F. *Teratology* **1974**, *9*, 11–16.
77. Robens, J. F. *Toxicol. Appl. Pharmacol.* **1969**, *15*, 152–163.
78. Smalley, H. E.; Curtis, J. M.; Earl, F. L. *Toxicol. Appl. Pharmacol.* **1968**, *13*, 392–403.
79. Budreau, C. H.; Singh, R. P. *Toxicol. Appl. Pharmacol.* **1973**, *24*, 324–332.
80. Tanimura, T.; Katsuya, T.; Nishimura, H. *Arch. Environ. Health* **1967**, *15*, 609–613.
81. Kimbrough, R. D.; Gaines, T. B. *Arch. Environ. Health* **1968**, *16*, 805–808.
82. Wang, G. M. *Teratogenesis Carcinog. Mutagen.* **1988**, *8*, 117–126.
83. Shane, B. S.; Dunn, H. O.; Kenney, R. M.; Hansel, W.; Visek, W. J. *Biol. Reprod.* **1969**, *1*, 41–48.
84. Nelson, B. K.; Brightwell, W. S.; Kriege, E. F. *Toxicol. Ind. Health* **1990**, *6*, 373–387.
85. Himmelberger, D. V.; Brown, B. W.; Cohen, E. N. *Am. J. Epidemiol.* **1978**, *108*, 670–679.
86. Axxelsson, G.; Rylander, R. *Int. J. Epidemiol.* **1984**, *13*, 94–98.
87. Carlo, G.; Hearn, S. *Prog. Clin. Biol. Res.* **1984**, *160*, 60.
88. Valanis, B. *Occup. Med. State Art Rev.* **1986**, *1*, 431–444.
89. Sewer, L. E.; Hessol, N. A. In *Reproduction: The New Frontier in Occupational and Environmental Health Resume*; Lockey, J. E.; Lemasters, G. K.; Keye, W. R., Eds.; Alan R. Liss: New York, 1984; pp 15–47.
90. Hogue, C. J. *Occup. Med. State Art Rev.* **1986**, *1*, 457–472.
91. Shane, B. S. *Environ. Sci. Technol.* **1989**, *23*, 1187–1195.
92. Nordstrom, S.; Beckman, L.; Nordenson, I. *Hereditas* **1978**, *88*, 43–46.
93. Nordstrom, S.; Beckman, L.; Nordenson, I. *Hereditas* **1978**, *88*, 51–54.
94. Nordstrom, S.; Beckman, L.; Nordenson, I. *Hereditas* **1979**, *90*, 291–296.
95. Cohen, E. N.; Bellville, J. W.; Brown, B. W. *Anesthesiology* **1971**, *35*, 343–347.
96. Knill-Jones, R. P.; Newman, B. J.; Spence, A. A. *Lancet* **1975**, *2*, 807–809.
97. Selevan, S. G.; Lindbohm, M.; Hornung, R. W.; Hemminki, K. *N. Engl. J. Med.* **1985**, *313*, 1173–1178.
98. Hemminki, K.; Mutanen, P.; Suloniemi, I.; Niemi M.-L.; Vanio, H. *Br. Med. J.* **1982**, *285*, 1461–1463.
99. Yoder, J.; Watson, M.; Benson, W. W. *Mutat. Res.* **1973**, *21*, 335–340.
100. Espir, M. L. E.; Hall, J. W.; Shirreffs, J. G.; Stevens, D. L. *Br. Med. J.* **1970**, *1*, 423–425.
101. Kiraly, J.; Szentesi, I.; Ruzicska, M. *Arch. Environ. Contam. Toxicol.* **1979**, *8*, 309–319.
102. Axxelsson, G.; Molin, I. *Int. J. Epidemiol.* **1988**, *17*, 363–369.
103. Kallen, B.; Thorbert, G. *Environ. Res.* **1985**, *37*, 313–339.
104. Rita, P.; Reddy, P. P.; Reddy, S. V. *Environ. Res.* **1987**, *44*, 1–5.
105. Gill, W. B.; Schumacher, G. F. B.; Bibbo, M. *J. Reprod. Med.* **1976**, *16*, 147–153.
106. Matsumoto, H. G.; Goyo K.; Takevchi, T. *J. Neuropathol. Exp. Neurol.* **1965**, *24*, 563–574.
107. Harada, Y. *Minamata Disease*; Kumamota University Study Group of Minamata Disease: Kumamoto, Japan, 1968; pp 73–91.
108. Snyder, R. D. *N. Engl. J. Med.* **1971**, *284*, 1014–1016.

109. Amin-Zaki, L.; Elhassani, S.; Majeed, M. A.; Clarkson, T. W.; Doherty, R. A.; Greenwood, M. *Pediatrics*, **1974**, *54*, 587–595.
110. Morton, M. S.; Elwood, P. C. *Teratology* **1974**, *10*, 318.
111. Frankenburg, F. R.; Lipinski, J. F. *N. Engl. J. Med.* **1983**, *309*, 311–312.
112. Weinstein, M. M.; Goldfield, M. *Am. J. Psychiatry* **1975**, *132*, 529–531.
113. Rogan, W. J.; Gladen, B. C.; Hung, K.; Koong, S.; Shih, L.; Taylor, J. S.; Wu, Y.; Yang, D.; Ragan, N. B.; Hsu, C. *Science (Washington, D.C.)* **1986**, *241*, 334–336.
114. Rogan, W. J. *Teratology* **1982**, *26*, 259–261.
115. Cutting, R. T.; Phuoc, T. H.; Ballo, J.; Benenson, M. W.; Evans, C. H. *Congenital Malformations, Hydatiform Moles and Stillbirths in the Republic of Vietnam 1960–1969*; U.S. Government Pub. No. 903-223; U.S. Government Printing Office: Washington, DC, 1970.
116. Meselsohn, M. S. *Proceedings of 13th Annual Meeting*; American Association for the Advancement of Science: Washington, DC, 1970; Herbicide Assessment Commission.
117. LaPorte, J. R. *Lancet* **1977**, *1*, 1049–1050.
118. Kundstadter, P. *A Study of Herbicides and Birth Defects in the Republic of Vietnam: An Analysis of Hospital Records*; National Academy of Sciences: Washington, DC, 1982.
119. Butler, N. R.; Goldstein, H.; Ross, E. M. *Br. Med. J.* **1972**, *2*, 127–130.
120. Harlap, S.; Shioni, P. *Lancet* **1980**, *2*, 173–176.
121. Lubs, M. L. E. *Am. J. Obstet. Gynecol.* **1973**, *115*, 66–76.
122. Mills, J. L.; Graubard, B. I.; Harley, E. E. *J. Am. Med. Assoc.* **1984**, *252*, 1875–1879.
123. Stessguth, A. P.; Carron, S. K.; Jones, K. L. *Lancet* **1985**, *2*, 285–291.
124. Khera, K. S. *CRC Crit. Rev. Toxicol.* **1987**, *17*, 345–375.
125. Happle, R.; Traupe, H.; Bounameaux Y.; Fitsh, T. *Dtsch. Med. Wochenschr.* **1984**, *109*, 1476–1480.
126. Whorton, M. D.; Krauss, R. M.; Marshall, S.; Milby, T. H. *Lancet* **1977**, *2*, 1259–1261.
127. Biava, C. G.; Smukler, E. A.; Whorton, M. D. *Exp. Mol. Pathol.* **1978**, *29*, 448–458.
128. Lipschultz, L. I.; Ross, C. E.; Whorton, M. D.; Milby, T. H.; Smith, R.; Joyner, R. E. *J. Urol.* **1980**, *12*, 464–468.
129. Whorton, M. D.; Foliart, D. E. *Mutat. Res.* **1983**, *123*, 13–30.
130. McLachlan, J. A.; Newbold, R. R.; Korach, K. S.; Lamb, J. C., IV; Suzuki, Y. In *Developmental Toxicology*; Kimmel, C. A.; Buelke-Sam, J., Eds.; Raven: New York, 1981; pp 213–232.
131. Lancranjan, I.; Popescu, H. I.; Gavanescu, O.; Klepsch, I.; Serbanescu, M. *Arch. Environ. Health* **1975**, *30*, 396–401.
132. Nordenson, I.; Beckman, G.; Beckman, L.; Nordstrom, S. *Hereditas* **1978**, *88*, 263–267.
133. Schardein, J. L.; *Chemically Induced Birth Defects*, Marcel Dekker: New York, 1985.
134. Fabro, S. *Fund. Appl. Toxicol.* **1985**, *5*, 609–614.
135. Jelovsek F. R.; Mattison, D. R.; Chen, J. J. *Obstet. Gynecol.* **1989**, *74*, 107–119.
136. Mattison, D. J.; Jelovsek, F. R. *Environ. Health Perspect.* **1987**, *76*, 107–119.

RECEIVED for review October 9, 1992. ACCEPTED revised manuscript April 8, 1993.

8

The Gene-Tox Program
Data Evaluation of Chemically Induced Mutagenicity

Michael C. Cimino and Angela E. Auletta

Office of Pollution Prevention and Toxics, Health and Environmental Review Division (7403), U.S. Environmental Protection Agency, Washington, DC 20460

> *The Gene-Tox Program of the U.S. Environmental Protection Agency is a multiphased effort to review and evaluate the existing literature in genetic toxicology (mutagenicity). In Phase I it selected assay systems for evaluation, generated expert panel reviews of the data from the scientific literature, and recommended testing protocols for the systems. Phase II established and evaluated the database for its relevance to identifying human health hazard. The ongoing Phase III continues reviewing the literature and updating chemical genetic toxicity data. Currently, data exist on over 4000 chemicals in 36 assay systems. The panel reports are published in the scientific literature, and the data are also publicly available through the National Library of Medicine TOXNET system. Public availability should increase Gene-Tox's utility, expand its analysis, and affect the manner and speed of its update.*

THE GENE-TOX PROGRAM of the U.S. Environmental Protection Agency (U.S. EPA) is a multiphased effort to review and evaluate the existing literature in chemically induced genetic toxicology. It has been an ongoing project under the Office of Pollution Prevention and Toxics (formerly the Office of Toxic Substances) since its inception in 1979. The first phase of the program, Phase I, was devoted to the selection of bioassays to be evaluated, the evaluation of literature by work groups of experts in each area, and the publication of reports of the resulting reviews. The second phase, Phase II, was devoted to the establishment of a database of chemicals evaluated by each work group and the analysis of that database. The last and current phase, Phase III (ongoing efforts), is devoted to the continued review of selected assays and the update of the database. Reports of all three phases are pub-

This chapter not subject to U.S. copyright
Published 1994 American Chemical Society

lished in the "Reviews in Genetic Toxicology" section of the journal *Mutation Research* (Appendix). Recently, the database has been made available on-line through the Toxicology Data Network system (TOXNET) of the National Library of Medicine (NLM).

Phase I

Work Groups. During Phase I of the Program, work groups of experts reviewed and evaluated the published literature for 36 selected assays (Table I) to determine the following for each system: Its validity; the chemicals for which it was best-suited (chemical-specific sensitivity of the system); its proper test protocol; and the appropriate techniques of data analysis, interpretation, and presentation.

In addition, each work group was asked to evaluate the assay's ability to discriminate between mutagens and nonmutagens or carcinogens and noncarcinogens; to evaluate the system's performance with chemicals of various classes and to identify chemicals whose effects were not adequately detected; to formulate generalized protocols and criteria for data evaluation and validation; to identify areas requiring additional research or further development and validation; and to publish an evaluation of the assay in the open literature.

Literature. Literature for evaluation was provided to the work groups by the Environmental Mutagen Information Center (EMIC), Oak Ridge National Laboratory, Oak Ridge, Tennessee. EMIC selected only the portion of the available literature that met the following criteria: The article was a primary paper published in a peer-reviewed journal; the article dealt with chemical mutagenesis, not with radiation-induced mutagenesis; the agent studied was a pure chemical; the article contained quantitative data; and the article was published in English or in a language for which EMIC had easy access to a translation.

Evaluation. The resulting articles were evaluated by the work groups for the following elements: proper use of experimental design; use of positive and negative controls; proper selection of solvents and vehicles; acceptable spontaneous background mutation frequency or rate; use of metabolic activation systems, if necessary, as for in vitro test systems; use of appropriate criteria for positive, negative, or inconclusive results; and provision of dose–response information. This last criterion was not considered critical if all others were met. In addition, each work group was free to apply other criteria that might be specific to its particular assay. Evaluated chemicals were designated as positive (+), positive with dose–response data provided (+ D), positive with metabolic activation only (+ *), negative (−), or evaluated but with no definitive conclusion (T).

Table I. Assays Evaluated in Phase I and Updated in Phase III, Number of Chemicals in Database for Each Assay (As of April 22, 1994), and Gene-Tox Report (See Appendix)

Assay System	Number of Chemicals	Report Number
Gene mutation		
Salmonella typhimurium[a]	2469	38
Escherichia coli[a]	158	3
Yeast	228	35
Fungi	56	34
Plants	221	18,19,24,25
Drosophila sex-linked recessive lethal[a]	420	32
Chinese hamster lung cells (V79)[a]	192	9
Chinese hamster ovary cells (CHO)[a]	121	5,43
Mouse lymphoma L5178Y cells[a]	45	30
Mouse spot test[a]	27	7
Mouse visible specific locus test[a]	22	6
Chromosomal effects		
Yeast	74	29
Fungi	25	12,13
Plant cytogenetics	289	17,20,21,22,23
Drosophila	70	36
Mammalian cytogenetics in vitro[a]	94	10
Mammalian cytogenetics in vivo[a]	44	10
Micronucleus assay[a]	417	31,44
Dominant lethal assay[a]	139	37
Mouse heritable translocation assay[a]	16	4
DNA damage and repair		
Repair-proficient and -deficient bacteria	526	11
Sister chromatid exchange in vitro[a]	642	8,45
Sister chromatid exchange in vivo	94	8,45
Unscheduled DNA synthesis	207	33
DNA repair	49	14
Oncogenic transformation		
Cell strains	115	26
Cell lines	163	26
Viral enhancement	222	26
Ancillary assays		
Host-mediated assay	207	15
Body fluid analysis	42	15
Sperm morphology	226	27,28

NOTE: The following assays will be added in Phase III; Drosophila SMART (somatic mutation and recombination test); Chinese hamster ovary cells (AS52); and mouse biochemical specific locus test.
[a]Phase I assays that are being updated in Phase III.

Results of Phase I. By the end of Phase I, the work groups had published 37 review articles, 36 concerned with assays in genetic toxicology and 1 describing the establishment of the Gene-Tox carcinogen database (1; see also Appendix).

Phase II

Fundamental Questions. In addition to the published reports, a computer database of over 2000 chemicals had been established at EMIC (2). At the outset of the Gene-Tox Program, it was anticipated that this database would be amenable to the type of analysis that would answer a series of fundamental questions in genetic toxicology:

- What genetic and related end points are of concern to human health?
- How can toxicologists distinguish test systems that are ready for extensive use in testing from those that should be regarded as still in a developmental stage?
- What is the sensitivity of each assay in responding to specific classes of chemicals, and what are the major strengths and weaknesses of each assay?
- What correlations exist between the mutagenesis and carcinogenesis end points?
- Is it possible to devise specialized batteries of bioassays that have high probabilities of detecting the various types of genetic and related damage induced by various classes of chemicals?
- Do in vitro mutagenesis and carcinogenesis bioassays provide comparable estimates of potency (strength of response)?
- What information gaps and future research needs can be identified, and what mechanisms can be established for continuing evaluation of the status of test systems?

Unfortunately, certain characteristics of the database have made such an analysis difficult, if not impossible, to perform.

Considerations. Four considerations are associated with the Gene-Tox database. The first is that there is no even distribution of chemicals across assay systems. For example, of the more than 1050 chemicals in the Phase I *Salmonella* database, approximately 200 (~19%) had also been tested in a cancer bioassay. In comparison, 59 of the approximately 200 chemicals in the mouse lymphoma L5178Y Phase I database (~30%) had been tested in a bioassay.

The second consideration is that there has proved to be little basis for comparative studies of mutagenesis assay systems. In the Phase I database, 1559 chemicals, or 59% of the total, had been tested in only 1 of the 36 systems. Chemicals that had been tested in more than one system were, for the most part, either direct-acting mutagens or those known to be metabolized to reactive intermediates by liver enzyme systems. This may have made the sensitivity of the various systems appear unnaturally high.

The third consideration is that the database is skewed to positive test results. With the exception of the *Salmonella* assay, there is a paucity of negative test results in the database in general and in the carcinogen database in particular, in which only 61 of 506 chemicals evaluated had negative results. This preponderance of positive results is a reflection of the interests of the original authors, for whom positive results hold more allure. It also reflects the greater ease in providing a conclusive positive result versus a conclusive negative one and the reluctance of scientific journals to publish seemingly less interesting negative data.

The fourth consideration is that the chemicals tested are unevenly distributed across the 30 chemical classes used in the Gene-Tox classification scheme (List 1). The most heavily tested classes are class 2 (acyl and aryl halides, halogenated ethers and halohydrins, and saturated and unsaturated alkyl halides), class 8 (aromatic amines, aliphatic amines, amides, and sulfonamides), class 25 (benzene rings), class 29 (alcohols and phenols), and class 30 (heterocyclic rings not otherwise classified and unclassified compounds). Such a distribution makes an analysis of chemical class specificity of the various assays difficult for all except the *Salmonella* assay, in which a sufficient number of chemicals have been tested in the various classes to permit a determination of system performance according to chemical class.

Results of Phase II. The Phase II analysis resulted in three publications on the establishment of the database (2), the evaluation of mutagenicity assays for the purpose of genetic risk assessment (3), and the developmental status of various assays for genetic toxicology testing (4; *see* also Appendix).

Phase III

Updating. As part of the ongoing Gene-Tox effort, certain assays from Phase I have been selected for update (*see* Table I). In addition, three assays not evaluated in Phase I will be added to the updated database: the Chinese hamster ovary (CHO) AS52 assay, the *Drosophila* somatic mutation and recombination tests (SMART), and the mouse biochemical specific locus test.

Although the update process has been simplified in comparison with that used in Phase I, the overall objectives of the program and the basic work-group structure remain in place. Over 1500 chemicals have been added to the database since the completion of Phase I, bringing the total number of chemicals evaluated to over 4000. The basic features of the Phase III database are the same as those noted for Phase I. There is still a paucity

List 1. Chemical Classification Scheme Used in Gene-Tox

Class	Chemicals Included
1	Acridines, quinacridines, and benzimidazoles
2	Acyl and aryl halides; halogenated ethers and halohydrins; and saturated and unsaturated alkyl halides
3	Aldehydes and anhydrides
4	Alkyl epoxides and aryl epoxides
5	Alkyl sulfates, sulfoxides, sulfones, sulfonates, and organic sulfur compounds not otherwise classified
6	Anthraquinones and quinones
7	Antibiotics and mycotoxins
8	Aromatic amines; aliphatic amines and amides and sulfonamides
9	Aziridines and nitrogen and sulfur mustards
10	Aromatic azo compounds, azoxy compounds, hydrazo compounds, diazoalkanes, nitriles, and azides
11	Carbamates, ureas, thioureas, and dicarboximides
12	Dioxins, xanthenes, thioxanthenes, and phenothiazines
13	Halogens and inorganic derivatives, and sulfur and nitrogen oxides
14	Hydrazides, hydrazines, and triazenes
15	Hydroxylamines and amine-N-oxides
16	Lactones and organic peroxides
17	Mineral fibers
18	Nitroimidazoles, nitrofurans, nitroquinolines, nitroaromatics, and nitroalkanes
19	Nitrosamides, nitrosoureas, and nitrosoguanidines
20	Nitrosamines
21	Organolead, organomercury, organophosphorus compounds, metals and derivatives, phosphoric acid esters, and phosphamides
22	Polycyclic aromatic hydrocarbons, fluorenones
23	Pyrimidine derivatives, and purine derivatives
24	Steroids
25	Benzene ring
26	Amino acids and derivatives
27	Alkaloids
28	Carbohydrates and derivatives
29	Alcohols and phenols
30	Heterocyclic rings not otherwise classified and unclassified compounds

of negative test results; the majority of the chemicals evaluated have been tested in only one system, and chemical class distribution is essentially unaffected.

Analysis of the Gene-Tox Database. The database for the *Salmonella* assay now totals 2469 chemicals (Table II). Of these, 326 have associated carcinogenicity data (Table III): 268 are classified as carcinogens, and 58 are classified as noncarcinogens. Of the 268 carcinogens, 210 are positive in *Salmonella,* and 58 are negative. Of the 58 noncarcinogens, 38 are negative in *Salmonella,* and 20 are positive. "Sensitivity" (the proportion of chemicals testing positive as both mutagens and carcinogens versus the total

Table II. The Gene-Tox *Salmonella* Database

Test System Result	Number of Chemicals Evaluated
Total number of chemicals evaluated	2469
Positive chemicals	1100
Positive without activation	666
Positive with activation	416
Positive without activation, negative with	18
Negative Chemicals	880
Chemicals with no definitive call	489

number that are carcinogens) is 78%; "specificity" (the proportion of tested chemicals that are both nonmutagenic and noncarcinogenic versus the total number that are noncarcinogenic) is 65%; and "accuracy" (concordance—the proportion of tested chemicals that are both mutagenic and carcinogenic or that are both nonmutagenic and noncarcinogenic, versus the total number of chemicals for which there are data in both systems) is 76%. "Positive predictivity" (the proportion of tested chemicals that are both carcinogenic and mutagenic compared with all mutagens in the assay) is 91%, and "negative predictivity" (the proportion of tested chemicals that are noncarcinogenic and nonmutagenic compared with all nonmutagens in the assay) is 39%.

Analysis of National Toxicology Program (NTP) Database. Evaluation of the results from the testing initiative of the NTP with 114 chemicals (5) produces figures different from those obtained from the Gene-Tox evaluation for the *Salmonella* assay: a sensitivity of 52%, a specificity of 91%, a concordance of 66%, positive predictivity of 90%, and negative predictivity of 55% (Table IV).

Comparison of Gene-Tox and NTP Databases. In considering these differences in test performances, we should observe several important differences between the Gene-Tox and NTP databases. Most obvious is the

Table III. Performance of *Salmonella* Assay (SAL) in Predicting Rodent Carcinogenicity (CCG)

Result	SAL Positive	SAL Negative	Total
CCG positive	210 (78% sensitivity) (91% positive predictivity)	58	268
CCG negative	20	38 (65% specificity) (39% negative predictivity)	58
Total	230	96	326

Table IV. Comparison of the Gene-Tox and NTP *Salmonella* Databases

Database	Sensitivity	Specificity	Accuracy (Concordance)	Positive Predictivity	Negative Predictivity
Gene-Tox	210/268 78%	38/58 65%	248/326 76%	210/230 91%	38/96 39%
NTP	35/67 52%	43/47 91%	75/114 66%	35/39 90%	43/78 55%

SOURCE: NTP data are taken from reference 5.

order of magnitude difference in the numbers of chemicals for which there are data in the two databases. In the case of the *Salmonella* assay, the system for which the greatest number of chemicals have been tested, more than 2400 chemicals have test data in Gene-Tox versus 114 for NTP. More importantly, the chemicals included in the NTP program were selected according to defined criteria and were tested according to prearranged and standardized protocols, whereas chemical selection in Gene-Tox database is more random (based on individual authors' interests), and the protocols are varied. However, in the case of the *Salmonella* assay, the most likely reason for the reported differences in sensitivity, specificity, predictive ability, and concordance is probably related to chemical class distribution of the agents tested.

The Gene-Tox chemical classification scheme is based on selected organic functional groups, ring systems, biological origins, and organic composition. Carcinogens that have been tested in the *Salmonella* assay are more apt to be classified as halides, epoxides, sulfur compounds, mustards, xanthenes, nitro and nitroso compounds, nitrosamines, metals, polycyclic aromatic hydrocarbons (PAHs), steroids, and benzene rings than would be expected if the chemical sampling were random.

Noncarcinogens in the Gene-Tox *Salmonella* database are primarily halides (class 2), aromatic amines (class 8), carbamates and ureas (class 11), nitro compounds (class 18), PAHs (class 22), and benzene ring compounds (class 25).

Usefulness of Mutagenicity Assays. On the basis of this analysis, it appears that the *Salmonella* assay can be a useful tool for identifying rodent carcinogens, as long as attention is paid to the importance of chemical class when results are interpreted.

Public Availability. The Gene-Tox database is now publicly available through the TOXNET system of the NLM. The TOXNET unit record for the Gene-Tox database is shown in List 2. This is the structure of the record as used by the NLM; it shows the abbreviated field names used in storing

List 2. TOXNET Gene-Tox Unit Record: Field Names and Descriptions

0.		** Administrative information
	GTN	Gene-Tox number (sequential order)
	DATE	last revision date
	RLEN	record length
	UPDT	update history
1.	ID	** Substance identification
	NAME	name of substance
	RN	CAS Registry Number
	SY	synonyms
	CCAT	chemical classification category
2.	MSTU	** Mutagenicity studies
	GENB	Gene-Tox evaluation B (post-1980)
		[species/cell type]
		[sex]
		[assay type]
		[assay code]
		[results]
		[activation]
		[dose response]
		[reference]
		[panel report]
	GENA	Gene-Tox evaluation a (pre-1980)
		[species/cell type]
		[sex]
		[assay type]
		[assay code]
		[results]
		[activation]
		[dose response]
		[reference]
		[panel report]

and searching for data and an expanded explanation of the contents of that field. An example of the data, a small portion of the unit record for formaldehyde, is shown in List 3.

In the future, updating the on-line database will primarily be the responsibility of the EPA with the assistance of EMIC. The update will continue to make use of the peer review system in a slightly modified form. Chemicals to be added to the database will continue to be published in *Mutation Research: Reviews in Genetic Toxicology;* when the manuscript is submitted, the data will simultaneously be added to the TOXNET file. Presently, this file contains all of the chemicals evaluated in Phase I plus the results of the Phase III updates for the CHO–hypoxanthine-guanine phosphoribosyltransferase, micronucleus and see assays.

List 3. Gene-Tox Unit Record: Sample (Partial Record of Formaldehyde)

GTN	- 14		
UPDT	- Complete update on 11/21/90, 6 fields added/edited/deleted		
RLEN	- 1593		
NAME	- formaldehyde		
RN	- 50-00-0		
GENB			
	Species/cell type	:	Chinese hamster ovary (CHO) cells
	Assay type	:	gene mutation at the HGPRT locus
	Assay code	:	CHOT
	Results	:	no conclusion
	Reference	:	EMIC/53976; *J. Toxicol. Environ. Health* **1983**, *12*; 27–38
	Panel report	:	EMIC/71517; *Mutat. Res.* **1988**, *196*, 17–36
GENB			
	Species/cell type	:	mammalian polychromatic erythrocytes, all species
	Assay	:	micronucleus test
	Assay code	:	MNTT
	Results	:	no conclusion
	Reference	:	EMIC/41641; *Mutat. Res.* **1981**, *90*, 91–109
	Panel report	:	EMIC/77345; *Mutat. Res.* **1990**, *239*, 29–80
GENA			
	Species/cell type	:	*Neurospora crassa*
	Assay type	:	reverse mutation
	Assay code	:	NER+
	Results	:	positive
	Panel report	:	EMIC/52327; *Mutat. Res.* **1984**, *133*, 87–134

Acknowledgments

The EPA acknowledges the many members of the genetic toxicology community who have so generously given their time and talent to contribute to the success of the program. We are also grateful to John Wassom and the staff of the EMIC, and to Dorothy Stroup and the staff of NLM for their unfailing support, without whom this program would not be possible.

References

1. Nesnow, S.; Argus, M.; Bergman, H.; Chu, K.; Frith, C.; Helmes, T.; McGaughy, R.; Ray, V.; Slaga, T. J.; Tennant, R.; Weisburger, E. *Mutat. Res.* **1987**, *185*, 1–195.
2. Ray, V. A.; Kier, L. D.; Kannan, K. L.; Haas, R. T.; Auletta, A. E.; Wassom, J. S.; Nesnow, S.; Waters, M.D. *Mutat. Res.* **1987**, *185*, 197–241.
3. Russell, L. B.; Aaron, C. S.; de Serres, F.; Generoso, W. M.; Kannan, K. L.; Shelby, M.; Springer, J.; Voytek, P. *Mutat. Res.* **1984**, *134*, 143–157.
4. Brusick, D.; Auletta, A. *Mutat. Res.* **1985**, *153*, 1–10.
5. Zeiger, E.; Haseman, J. K.; Shelby, M. D.; Margolin, B. H.; Tennant, R. W. *Environ. Mol. Mutagen.* **1990**, *16*(Suppl. 18), 1–14.

Appendix
List of Publications[1] of the Gene-Tox Program of the U.S. EPA

Introductory Papers
1. Green, S.; Auletta, A. "Editorial Introduction to the Reports of 'The Gene-Tox Program'. An Evaluation of Bioassays in Genetic Toxicology". *Mutat. Res.* **1980**, *76*, 165–168.
2. Waters, M. D.; Auletta, A. "The Gene-Tox Program: Genetic Activity Evaluation". *J Chem. Inf. Comp. Sci.* **1981**, *21*, 35–38.

Phase I Panel Reports
3. Brusick, D. J.; Simmon, V. F.; Rosenkranz, H. S.; Ray, V. A.; Stafford, R. S. "An Evaluation of *Escherichia coli* WP2 and WP2 *uvrA* Reverse Mutation Assay". *Mutat. Res.* **1980**, *76*, 169–190.
4. Generoso, W. M.; Bishop, J. B.; Gosslee, D. G.; Newell, G. W.; Sheu, C.-J.; Von Halle, E. "Heritable Translocation Test in Mice". *Mutat. Res.* **1980**, *76*, 191–215.
5. Hsie, A. W.; Casciano, D. A.; Couch, D. B.; Krahn, D. F.; O'Neill, J. P.; Whitfield, B. L. "The Use of Chinese Hamster Ovary Cells To Quantify Specific Locus Mutation and To Determine Mutagenicity of Chemicals. A Report of the Gene-Tox Program". *Mutat. Res.* **1981**, *86*, 193–214.
6. Russell, L. B.; Selby, P. B.; Von Halle, E.; Sheridan, W.; Valcovic, L. "The Mouse Specific-Locus Test with Agents Other Than Radiations. Interpretation of Data and Recommendations for Future Work". *Mutat. Res.* **1981**, *86*, 329–354.
7. Russell, L. B.; Selby, P. B.; Von Halle, E.; Sheridan, W.; Valcovic, L. "Use of the Mouse Spot Test in Chemical Mutagenesis: Interpretaions of Past Data and Recommendations for Future Work". *Mutat. Res.* **1981**, *86*, 355–379.
8. Latt, S. A.; Allen, J.; Bloom, S. E.; Carrano, A.; Falke, E.; Kram, D.; Schneider, E.; Schreck, R.; Tice, R.; Whitfield, B.; Wolff, S. "Sister-Chromatid Exchanges: A Report of the Gene-Tox Program". *Mutat. Res.* **1981**, *87*, 17–62.
9. Bradley, M. O.; Bhuyan, B.; Francis, M. C.; Langenbach, R.; Peterson, A.; Huberman, E. "Mutagenesis by Chemical Agents in V79 Chinese Hamster Cells: A Review and Anal-

[1]Reprints still available may be obtained from John S. Wassom, Environmental Mutagen Information Center, Oak Ridge National Laboratory, Building 2001 Mail Stop 6050, P.O. Box 2008, Oak Ridge, TN 37831.

ysis of the Literature. A Report of the Gene-Tox Program". *Mutat. Res.* **1981**, *87*, 81–142.
10. Preston, R. J.; Au, W.; Bender, M. A.; Brewen, J. G.; Carrano, A. V.; Heddle, J. A.; McFee, A. F.; Wolff, S.; Wassom, J. S. "Mammalian In Vivo and In Vitro Cytogenetic Assays: A report of the U.S. EPA's Gene-Tox Program". *Mutat. Res.* **1981**, *87*, 143–188.
11. Leifer, Z.; Kada, T.; Mandel, M.; Zeiger, E.; Stafford, R.; Rosenkranz, H. S. "An Evaluation of Tests Using DNA Repair-Deficient Bacteria for Predicting Genotoxicity and Carcinogenicity. A Report of the U.S. EPA's Gene-Tox Program". *Mutat. Res.* **1981**, *87*, 211–297.
12. Kafer, E.; Scott, B. R.; Dorn, G. L.; Stafford, R. "*Aspergillus nidulans:* Systems and Results of Tests for Chemical Induction of Mitotic Segregation and Mutation. I. Diploid and Duplication Assay Systems. A Report of the U.S. EPA Gene-Tox Program". *Mutat. Res.* **1982**, *98*, 1–48.
13. Scott, B. R.; Dorn, G. L.; Kafer, E.; Stafford, R. "*Aspergillus nidulans:* Systems and Results of Tests for Induction of Mitotic Segregation and Mutation. II. Haploid Assay Systems and Overall Response of All Systems. A Report of the U.S. EPA Gene-Tox Program". *Mutat. Res.* **1982**, *98*, 49–94.
14. Larsen, K. H.; Brash, D.; Cleaver, J. E.; Hart, R. W.; Maher, V. M.; Painter, R. B.; Sega, G. A. "DNA Repair Assays as Tests for Environmental Mutagens. A Report of the U.S. EPA Gene-Tox Program". *Mutat. Res.* **1982**, *98*, 287–318.
15. Legator, M. S.; Bueding, E.; Batzinger, R.; Conner, T. H.; Eisenstadt, E.; Farrow, M. G.; Fiscor, G.; Hsie, A.; Seed, J.; Stafford, R. S. "An Evaluation of the Host-Mediated Assay and Body Fluid Analysis. A Report of the U.S. Environmental Protection Agency Gene-Tox Program". *Mutat. Res.* **1982**, *98*, 319–374.
16. Constantin, M. J.; Owens, E. T. "Introduction and Perspectives of Plant Genetic and Cytogenetic Assays. A Report of the U.S. Environmental Protection Agency Gene-Tox Program". *Mutat. Res.* **1982**, *99*, 1–12.
17. Constantin, M. J.; Nilan, R. A. "Chromosome Aberration Assays in Barley *(Hordeum vulgare).* A Report of the U.S. Environmental Protection Agency Gene-Tox Program". *Mutat. Res.* **1982**, *99*, 13–36.
18. Constantin, M. J.; Nilan, R. A. "The Chlorophyll-Deficient Mutant Assay in Barley *(Hordeum vulgare).* A Report of the

U.S. Environmental Protection Agency Gene-Tox Program". *Mutat. Res.* **1982**, *99*, 36–49.
19. Redei, G. P. "Mutagen Assay in Arabidopsis. A Report of the U.S. Environmental Protection Agency Gene-Tox Program". *Mutat. Res.* **1982**, *99*, 243–255.
20. Ma, T.-H. "*Vicia* Cytogenetic Tests for Environmental Mutagens. A Report of the U.S. Environmental Protection Agency Gene-Tox Program". *Mutat. Res.* **1982**, *99*, 257–271.
21. Grant, W. F. "Chromosome Aberration Assays in Allium. A Report of the U.S. Environmental Protection Agency Gene-Tox Program. *Mutat. Res.* **1982**, *99*, 273–291.
22. Ma, T.-H. "Tradescantia Cytogenetic Tests (Root-Tip Mitosis, Pollen Mitosis, Pollen Mother Cell Meiosis). A Report of the U.S. Environmental Protection Agency Gene-Tox Program". *Mutat. Res.* **1982**, *99*, 293–302.
23. Van't Hof, J.; Schairer, L. A. "Tradescantia Assay System for Gaseous Mutagens. A Report of the U.S. Environmental Protection Agency Gene-Tox Program". *Mutat. Res.* **1982**, *99*, 303–315.
24. Plewa, M. J. "Specific-Locus Mutation Assays in *Zea mays*. A Report of the U.S. Environmental Protection Agency Gene-Tox Program". *Mutat. Res.* **1982**, *99*, 317–337.
25. Vig, B. K. "Soybean (*Glycine max* [L.] merrill) as a Short-Term Assay for the Study of Environmental Mutagens. A Report of the U.S. Environmental Protection Agency Gene-Tox Program". *Mutat. Res.* **1982**, *99*, 339–347.
26. Heidelberger, C.; Freeman, A. E.; Pienta, R. J.; Sivak, A.; Bertram, J. S.; Casto, B. C.; Dunkel, V. C.; Francis, M. W.; Kakunaga, T.; Little, J. B.; Schechtman, L. M. "Cell Transformation by Chemical Agents—A Review and Analysis of the Literature. A Report of the U.S. Environmental Protection Agency Gene-Tox Program". *Mutat. Res.* **1983**, *114*, 283–385.
27. Wyrobek, A. J.; Gordon, L. A.; Burkhart, J. G.; Francis, M. W.; Kapp, R. W., Jr.; Letz, G.; Malling, H. V.; Topham, J. C.; Whorton, M. D. "An Evaluation of the Mouse Sperm Morphology Test and Other Sperm Tests in Nonhuman Mammals. A Report of the U.S. Environmental Protection Agency Gene-Tox Program". *Mutat. Res.* **1983**, *115*, 1–72.
28. Wyrobek, A. J.; Gordon, L. A.; Burkhart, J. G.; Francis, M. W.; Kapp, R. W., Jr.; Letz, G.; Malling, H. V.; Topham, J. C.; Whorton, M. D. "An Evaluation of Human Sperm as Indicators of Chemically Induced Alterations of Spermato-

genic Function. A Report of the U.S. Environmental Protection Agency Gene-Tox Program". *Mutat. Res.* **1983**, *115*, 73–148.

29. Loprieno, N.; Barale, R.; Von Halle, E. S.; von Borstel, R. C. "Testing of Chemicals for Mutagenic Activity with *Schizosaccharomyces pombe*. A Report of the U.S. Environmental Protection Agency Gene-Tox Program". *Mutat. Res.* **1983**, *115*, 215–223.

30. Clive, D.; McCuen, R.; Spector, J. F. S.; Piper, C.; Mavournin, K. H. "Specific Gene Mutations in L5178Y Cells in Culture. A Report of the U.S. Environmental Protection Agency Gene-Tox Program". *Mutat Res.* **1983**, *115*, 225–251.

31. Heddle, J. A.; Hite, M.; Kirkhart, B.; Mavournin, K.; MacGregor, J. T.; Newell, G. W.; Salamone, M. F. "The Induction of Micronuclei as a Measure of Genotoxicity. A Report of the U.S. Environmental Protection Agency Gene-Tox Program". *Mutat. Res.* **1983**, *123*, 61–118.

32. Lee, W. R.; Abrahamson, S.; Valencia, R.; Von Halle, E. S.; Wurgler, F. E.; Zimmering, S. "The Sex-Linked Recessive Lethal Test for Mutagenesis in *Drosophila melanogaster*. A Report of the U.S. Environmental Protection Agency Gene-Tox Program". *Mutat. Res.* **1983**, *123*, 183–279.

33. Mitchell, A. D.; Casciano, D. A.; Meltz, M. L.; Robinson, D. E.; San, R. H. C.; Williams, G. M.; Von Halle, E. S. "Unscheduled DNA Synthesis Tests. A Report of the U.S. Environmental Protection Agency Gene-Tox Program". *Mutat. Res.* **1983**, *123*, 363–410.

34. Brockman, H. E.; de Serres, F. J.; Ong, T.-m.; DeMarini, D. M.; Katz. A. J.; Griffiths, A. J. F.; Stafford, R. S. "Mutation Tests in *Neurospora crassa*. A Report of the U.S. Environmental Protection Agency Gene-Tox Program". *Mutat. Res.* **1984**, *133*, 87–134.

35. Zimmernman, F. K.; von Borstel, R. C.; Von Halle, E. S.; Parry, J. M.; Siebert, D.; Zetterberg, G.; Barale, R.; Loprieno, N. "Testing of Chemicals for Genetic Activity with *Saccharomyces cerevisiae*: A Report of the U.S. Environmental Protection Agency Gene-Tox Program". *Mutat. Res.* **1984**, *133*, 199–244.

36. Valencia, R.; Abrahamson, S.; Lee, W. R.; Von Halle, E. S.; Woodruff, R. C.; Wurgler, F. E.; Zimmering, S. "Chromosome Mutation Tests for Mutagenesis in *Drosophila melanogaster*. A Report of the U.S. Environmental Protection Agency Gene-Tox Program". *Mutat. Res.* **1984**, *134*, 61–88.

37. Green, S.; Auletta, A.; Fabricant, J.; Kapp, R.; Manandhar, M.; Sheu, C.-j.; Springer, J.; Whitfield, B. "Current Status of Bioassays in Genetic Toxicology—The Dominant Lethal Assay. A Report of the U.S. Environmental Protection Agency Gene-Tox Program". *Mutat. Res.* **1985**, *154*, 49–67.
38. Kier, L. D.; Brusick, D. J.; Auletta, A. E.; Von Halle, E. S.; Brown, M. M.; Simmon, V. F.; Dunkel, V.; McCann, J.; Mortlemans, K.; Prival, M.; Rao, T. K.; Ray, V. "The *Salmonella* typhimurium/Mammalian Microsomal Assay. A Report of the U.S. Environmental Protection Agency Gene-Tox Program". *Mutat. Res.* **1986**, *168*, 69–240.
39. Nesnow, S.; Argus, M.; Bergman, H.; Chu, K.; Frith, C.; Helmes, T.; McGaughy, R.; Ray, V.; Slaga, T. J.; Tennant, R.; Weisburger, E. "Chemical Carcinogens: A Review and Analysis of the Literature of Selected Chemicals and the Establishment of the Gene-Tox Carcinogen Data Base". *Mutat. Res.* **1987**, *185*, 1–195.

Phase II Assessment Panel Reports
40. Russell, L. B.; Aaron, C. S.; de Serres, F.; Generoso, W. M.; Kannan, K. L.; Shelby, M.; Springer, J.; Voytek, P. "A Report of the U.S. Environmental Protection Agency Gene-Tox Program. Evaluation of Mutagenicity Assays for Purposes of Genetic Risk Assessment". *Mutat. Res.* **1984**, *134*, 143–157.
41. Brusick, D.; Auletta, A. "Developmental Status of Bioassays in Genetic Toxicology. A Report of Phase II of the U.S. Environmental Protection Agency Gene-Tox Program". *Mutat. Res.* **1985**, *153*, 1–10.
42. Ray, V. A.; Kier, L. D.; Kannan, K. L.; Haas, R. T.; Auletta, A. E.; Wassom, J. S.; Nesnow, S.; Waters, M. D. "An Approach To Identifying Specialized Batteries of Bioassays for Specific Classes of Chemicals: Class Analysis Using Mutagenicity and Carcinogenicity Relationships and Phylogenetic Concordance and Discordance Patterns. 1. Composition and Analysis of the Overall Data Base. A Report of Phase II of the U.S. Environmental Protection Agency Gene-Tox Program". *Mutat. Res.* **1987**, *185*, 197–241.

Phase III Panel Reports
43. Li, A. P.; Gupta, R. S.; Heflich, R. H.; Wassom, J. S. "A Review and Analysis of the Chinese Hamster Ovary/Hypoxanthine Guanine Phosphoribosyl Transferase Assay To Determine the Mutagenicity of Chemical Agents. A Report of

Phase III of the U.S. Environmental Protection Agency Gene-Tox Program". *Mutat. Res.* **1988**, *196*, 17–36.
44. Mavournin, K. H.; Blakey, D. H.; Cimino, M. C.; Salamone, M. F.; Heddle, J. A. "The In Vivo Micronucleus Assay in Mammalian Bone Marrow and Peripheral Blood. A Report of the U.S. Environmental Protection Agency Gene-Tox Program". *Mutat. Res.* **1990**, *239*, 29–80.
45. Tucker, J. D.; Auletta, A.; Cimino, M. C.; Dearfield, K. L.; Jacobson-Kram, D.; Tice, R. R.; Caranno, A. V. "Sister-Chromatid Exchange: Second Report of the Gene-Tox Program". *Mutat. Res.* **1993**, *279*, 101–180.

RECEIVED for review September 3, 1992. ACCEPTED revised manuscript February 17, 1993.

9

Biological, Biochemical, and Molecular Markers in Environmental Epidemiology

Marilyn F. Vine

Department of Epidemiology, School of Public Health, University of North Carolina, Chapel Hill, NC 27599-7400

> *A biological or molecular marker is a cellular, biochemical, or molecular alteration that is measurable in biological media, such as human tissues, cells, or fluids. Biological markers are being explored to determine their ability to help us understand the intermediate steps between exposure and disease occurrence and to improve the precision of exposure and outcome measures in epidemiological studies. The purpose of this chapter is to discuss (1) the potential benefits of using biological markers in epidemiological research, (2) the chief characteristics of biological markers that make them particularly useful in epidemiological studies, (3) the limitations of the use of biological markers in epidemiological research, and (4) the rewards and obstacles involved in collaborative efforts between laboratory scientists, who develop the assays to detect markers, and epidemiologists.*

EPIDEMIOLOGY IS THE STUDY of the distribution and determinants of diseases in populations. Traditionally, epidemiologists have taken a "black box" approach to the study of environmentally induced disease, assessing an environmental or external exposure by using, for example, questionnaires and environmental monitoring and then evaluating the association between the exposure measure and some health outcome. Biological markers are being explored to determine their ability to help us understand the intermediate steps between exposure and disease occurrence and to improve the precision of exposure and outcome measures.

This chapter provides an overview of biological markers used in epidemiological studies. It discusses the properties of markers that make them potentially valuable to epidemiologists, and reviews criteria for the selection of markers to be used in epidemiological investigations. The current limitations of markers in epidemiological research are also discussed. Because laboratory assays are involved in measuring the presence of biological markers in human biological material, the benefits and challenges of collaborative efforts between epidemiologists and laboratory scientists are presented. Examples of situations in which biological markers have been used in environmental–occupational investigations are provided.

Types of Biological Markers

A biological marker is a "cellular, biochemical, or molecular alteration that is measurable in biological media such as human tissues, cells or fluids" (1). Biological markers can be measured in a variety of target and surrogate tissues. Target tissues are tissues that undergo pathological changes as the result of exposure (2). An example of a target tissue for cigarette smoke exposure is the lung. Marker levels in surrogate (nontarget) tissues, such as blood or urine, can sometimes be used to estimate marker levels in target tissues.

Biological markers can be conceptualized as forming a continuum that represents a sequence of events, from exposure to an exogenous agent to disease occurrence (3). Figure 1 (4) presents the categories of biological markers that represent the various stages in the sequence of events. These categories include markers of internal dose, biologically effective dose, early response, altered structure and function, and disease. This sequence of events begins when an individual is exposed to some environmental or external agent and is believed to progress, unless the agent is eliminated from the body and the damage that results from the exposure is repaired. Outside of the sequence of events are susceptibility factors that can act at any point along the continuum to modify the effects of external exposures. This conceptual framework is based on the work of Perera and Weinstein (5).

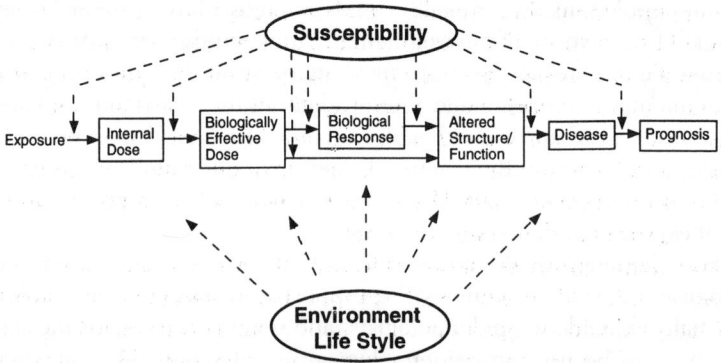

Figure 1. The relationship of biological markers to exposure and disease. (Modified with permission from reference 4. Copyright 1989 National Academy Press.)

A marker of internal dose indicates that an exogenous chemical has entered the body of an exposed individual. An exogenous chemical may enter the body by inhalation, ingestion, or absorption through the skin. The internal dose marker may measure the chemical itself (unchanged) or a metabolically altered state (3). An example of an internal dose marker that can be measured unchanged is lead in blood (6). A marker of a metabolically altered chemical is cotinine in the body fluids of smokers. Cotinine, a metabolic byproduct of nicotine metabolism (7), is a marker of exposure to tobacco smoke.

A marker of biologically effective dose (BED) indicates "the amount of absorbed chemical that has interacted with critical subcellular targets, measured in either a target or a surrogate tissue" (3). Workers exposed to coke oven emissions containing polycyclic aromatic hydrocarbons are at increased risk of lung cancer (8). DNA adducts (complexes of chemicals covalently bonded to DNA, derived from lung tissue, and obtained at autopsy or through biopsy) could potentially serve as a biologically effective dose marker for the target tissue. DNA adducts are believed to be involved in the initiation of cancer (9). Although it is unlikely that hemoglobin adducts (complexes of electrophilic chemicals covalently bonded with hemoglobin in red blood cells) are directly involved in the carcinogenic process, they may serve as surrogate BED markers of DNA adduct formation in target tissues. Animal (10–12) and human studies (13) suggest that for some chemicals, an increase in hemoglobin adduct formation is associated with an increase in DNA adduct formation in target tissues.

A marker of biologic response indicates "biological or biochemical changes in target cells or tissues that result from the action of the chemical and are thought to be a step in the pathologic process toward disease" (1). Examples of markers of biological response include chromosomal aberrations and micronuclei. For example, betel quid chewers who have an elevated risk of oral cancer also have elevated levels of micronuclei in exfoliated buccal mucosa cells (14). Micronuclei, small pieces of DNA that arise in the cytoplasm when chromatid or chromosomal fragments or whole chromosomes are not incorporated into daughter nuclei during mitosis (15), form as a result of exposure to clastogenic agents, such as X-rays, and agents that cause damage to the cell's spindle apparatus (16).

A marker of disease indicates a "biological or chemical event that either represents a subclinical stage of disease or is a manifestation of the disease itself" (1). Disease markers are often the dependent, or outcome, variables in epidemiological research. An example of a disease marker is serum α-fetoprotein, an indicator of liver cancer (1).

A marker of susceptibility "measures or is associated with, factors that increase an individual's risk of developing a disease after exposure to some exogenous agent" (17). Markers of susceptibility may be genetically determined or acquired. Xeroderma pigmentosum is an example of a genetically determined susceptibility. Individuals with this condition are at increased risk of developing skin cancer after exposure to ultraviolet radiation (18). Other examples of genetically determined susceptibilities include debrisoquin hydroxylation phenotype and N-acetylation phenotype. Extensive metabolizers of debrisoquin are thought to be at increased risk of lung cancer on exposure to cigarette smoke (19), and slow acetylators are at increased risk of bladder cancer on exposure to N-substituted aryl compounds (20), such as those used in the dye industry. Examples of acquired susceptibility factors include age, diet, life style, and previous diseases. For example, individuals who consume large quantities of alcohol and also smoke are at increased risk of dying from head and neck cancers (21), and people who have been infected with the hepatitis B virus have been found to have a greater chance of developing liver cancer in association with exposure to aflatoxin (22).

Advantages of Using Biological Markers in Epidemiological Research

There are potentially many reasons for using biological markers in epidemiological research (1,23):

1. Internal dose and biologically effective dose markers may provide more accurate measures of exposure that are specific to the individual than do traditional measures of exposure, such as area air monitoring, information about job title, or questionnaire data. In addition, biologically effective dose markers may provide estimates of the relevant exposure dose to the target tissue. By providing exposure information specific to the individual (and to target tissues, in particular), markers may help reduce misclassification in exposure measures, thereby enhancing our ability to detect dose–response relationships between external exposures and health outcome measures.
2. Biological markers may improve knowledge of participant compliance with treatment regimens in intervention trials. In a typical trial, one group of participants is randomized to receive one treatment or intervention, and one or more other groups are randomized to receive another treatment or a placebo. For example, in a study of the effects of β-carotene supplementation on micronuclei induction in buccal cells of betel quid chewers over a 3-month period, one group of participants would be given β-carotene capsules, and the other group would be given a placebo. Analysis of blood specimens several times during the 3-month trial would indicate whether the treatment group was really taking the β-carotene capsules

and whether the serum β-carotene levels of treated and control groups actually differed. It is possible, for example, that members of the control group were also taking vitamin supplements or consuming diets high in β-carotene.

3. Markers of disease may provide evidence of preclinical disease or may provide a more homogeneous classification of disease. Markers of preclinical disease may allow detection at earlier preventable stages. Markers that classify disease into more homogeneous subclassifications may help identify exposure–disease associations. Certain leukemias, for example, can be distinguished on the basis of whether oncogene activation is associated with the disease. Taylor et al. (24) reported that patients occupationally exposed to certain chemicals are more likely to develop *ras*-positive than *ras*-negative acute myelogenous leukemia (AML). Comparisons with healthy controls showed that occupational chemical exposure did not increase the risk of *ras*-negative AML.

4. Biological markers may provide clues to mechanisms of disease causation by elucidating intermediate steps in the process. Markers of response may be used to determine whether exogenous exposures have adverse effects on genetic material, an intermediate step in the development of cancer. For example, occupational exposure to ethylene oxide leads to an increased incidence of sister chromatid exchanges among exposed workers (25).

5. Markers of susceptibility may indicate subpopulations at increased risk of disease. If populations at increased risk can be identified, then these groups can be targeted for preventive measures.

6. Markers of exposure may improve knowledge of the extent of population exposures to various exogenous agents by means of surveillance techniques. Tissue specimens (e.g., blood or fat), obtained from appropriately sampled members of the population, can be analyzed to determine the extent of exposure to specific exogenous environmental agents in the population.

7. Markers of exposure may be used to validate more traditional measures of exposure. For example, in a study of the consumption of polychlorinated biphenyl (PCB)-contaminated fish and adverse reproductive effects among women in Wisconsin (26), serum PCB levels were used to assess the validity of using fish consumption (obtained via questionnaire) as an estimate of PCB intake in a subgroup of the women. The correlation between the questionnaire information and serum

PCB levels was 0.666. It was concluded that the questionnaire information, which was easier to obtain and less costly than measuring serum PCB levels, provided a reasonably accurate estimate of PCB intake (26).
8. Markers may provide more accurate measures of potential confounders in epidemiological research. Misclassification of confounder status can result in inadequate control of confounding factors in the analysis of epidemiological studies, potentially leading to a bias in study results (27).

For an expanded discussion of the potential benefits of using biological markers in epidemiological research mentioned, see Hulka, 1990 (1) and 1991 (23).

Criteria for Selecting Biological Markers in Epidemiological Research

An epidemiologist must consider many criteria before employing a biological marker in an epidemiological investigation. The most important consideration is the study's objective. If it is determined that a marker will benefit the study, the epidemiologist must determine which marker or markers will best meet the study's objective. The following characteristics of markers should be considered before a marker is used in an epidemiological investigation.

Availability of Markers. If only one marker exists to measure a particular exposure or outcome event, then the choice of a marker is simplified. Sometimes, however, more than one marker is available. For example, there are many potential markers of tobacco smoke exposure, including thiocyanate, carboxyhemoglobin, nicotine, cotinine, DNA and hemoglobin adducts, and urine mutagenicity. Each marker has different properties. The choice of marker depends on how well the properties of the available markers coincide with the objectives of the study.

Specificity of the Marker. Specificity refers to the ability of the marker to indicate the specific exposure or outcome of interest. For example, the urine mutagenesis assay is a nonspecific marker of mutagenic agents in the urine. It cannot indicate which exposure led to mutagens appearing in the urine. On the other hand, cotinine (metabolite of nicotine) in the urine is a marker essentially specific to tobacco smoke. The required degree of specificity depends on the research question.

Invasiveness of the Technique To Obtain Biological Specimens. Marker selection is often influenced by the quantity of the specimen needed to measure the marker and the invasiveness (either physically or psychologically) of the sampling technique necessary to obtain the biological specimen in which the marker is measured. The more invasive the procedure and the greater the amount of specimen needed, the less likely it is that people will participate in the study. Assays, for example, that detect micronucleus formation can be performed on human bone marrow cells, lymphocytes, red blood cells, and exfoliated cells. It is unlikely that many people would participate in a study in which sternal puncture to obtain bone marrow cells was required. Participation would be greater in a study that required the drawing of blood or the scraping of the oral mucosa to obtain exfoliated cells. The invasiveness of the procedures affects not only the numbers of participants but also their characteristics. Self-selection for a study on the basis of the technique used to obtain specimens could limit the generalizability of results; it may even lead to a bias in results because of systematic differences among outcome measures of participants, on the basis of exposure status.

Time to Appearance of the Marker. The "time to appearance of the marker" refers to the first point after the event of interest occurs at which the marker can be measured. In other words, there may be a delay between the time of the event of interest and when the marker indicates that it has occurred. For example, in micronucleus formation, micronuclei that are sloughed (exfoliated) from the surface of the buccal mucosa indicate damage that has occurred in the basal cell layer (14). Exfoliated cells are derived from basal cells. Only the basal cells divide and therefore form micronuclei. Stich et al. (28) have estimated that it takes 5–7 days for radiation-induced micronucleated basal cells to migrate to the surface of the buccal mucosa, where they are exfoliated. Thus, evaluating buccal mucosa cells sampled before the 5–7 days would underestimate the effect of the exposure on micronucleus formation. The time to appearance of the marker is often a greater concern in acute versus chronic exposure situations and when one is trying to link the time of exposure to a particular effect.

Persistence of the Marker. The persistence of the marker also influences the time during which one can evaluate the presence of the marker in a biological medium. The persistence of the marker depends on the metabolism, storage, and excretion of the marker from the body and also on the stability of the biological tissue in which the marker is measured. Cotinine, for example, has a half-life in serum of about 16–19 h (29), whereas nicotine, the parent compound, has a half-life of only 2 h (30). Nicotine levels would indicate short-term exposures, whereas cotinine levels would indicate exposures that took place over the previous 1 or 2 days. Hemoglobin adducts, however, could be used to indicate exposures over a longer period of time, because they are thought to persist for the life of the red blood cell, about 120 days (31).

Peak or Integrated Dose. Whether the level of the measured marker represents the peak or integrated dose over time depends on (1) whether the event that produced the marker was an acute (one time) or chronic (continuous) event, and (2) the tissue in which the marker is measured. Marker levels indicative of acute events, such as exposure to an exogenous chemical, rise (peak) and fall with time as the body metabolizes, excretes, or stores the chemical in another tissue. Thus, in acute events, when one samples the tissue in relation to the time of the event is important. If the tissue in which the marker is measured does not store the marker, chronic events often lead to a steady state of marker levels, in which similar levels of the marker enter and leave the tissue. If the tissue stores the chemical, one may be able to measure the accumulated dose of the marker. Lead, for example, accumulates in bone. Bone lead levels account for about 95% of the total body burden of lead. Lead has a half-life in bone of about a decade (6). Therefore, analysis of bone lead levels should indicate accumulated lead exposure.

Variability (Intraperson, Interperson). Variability is a major issue in biomarker research. Levels of markers can vary within a person (intraperson variability) from one time to the next and even among tissues within the same person. Levels of markers can also vary dramatically among individuals (interperson variability). Interperson variability in marker levels can result from differences in exogenous exposures as well as genetic variability in DNA repair processes and metabolism of exogenous agents. For example, 4-aminobiphenyl (4-ABP) hemoglobin adduct levels among active and passive smokers vary with acetylation and oxidation phenotype (32). It is necessary to understand the sources of variability in biological marker research in order to be able to make valid comparisons of marker levels among individuals. For a more extensive discussion of the sources of variability in epidemiological studies involving biological markers, *see* Hulka and Margolin (33).

Stability of the Marker in Storage. Stability of the marker in storage influences the feasibility of an epidemiological investigation. Storage procedures can include freezing or fixing tissue specimens in a preservative such as formaldehyde. If the marker is not stable in storage, analyses often have to be done concurrently with sample collection. Sometimes this procedure is not feasible. Sometimes having to analyze specimens at the time of sample collection escalates costs. For example, someone has to be employed to do the analyses throughout the study. If analyses are done over time, it is essential to analyze simultaneously specimens from control and exposed or diseased persons in order to control for any changes in laboratory practice over time. If the marker is stable in storage, analyses can be performed after all the samples have been collected, thereby increasing study design options.

Suitability of the Assay for an Epidemiological Study. The key issues in evaluating the suitability of an assay for an epidemiological study include the availability and cost of the assay as well as its sensitivity, specificity, and reliability. Some assays are commercially available, whereas others are either very complicated or still in development, such that only one or two laboratories in the world perform them. The success of epidemiological studies involving assays performed by few laboratory investigators depends on the epidemiologist's ability to collaborate with the particular laboratory investigators who perform the assay.

Costs of assays vary with how time-consuming and labor-intensive their procedures are. Analyses to determine serum cotinine levels, for example, are much less expensive than DNA adduct analyses. High assay costs can either rule out an assay for a particular application or limit the study's sample size.

Sensitivity of the assay must be considered. The assay must be able to detect population levels of the marker. In addition to being sensitive, it should be as specific as possible for the marker in question. For example, one criticism of radioimmunoassay techniques is that the antibodies used to detect a particular marker (antigen) can cross-react with similar compounds. A nonspecific assay can lead to misclassification of marker levels. Similarly, an unreliable assay (one that does not produce repeatable results on the same specimen) can also lead to misclassification of marker levels.

For a more in-depth discussion of the properties of markers and the criteria used in marker selection, see Wilcosky and Griffith (34) and Wilcosky (2).

Limitations of Biological Markers in Epidemiological Research

There are several limitations of the use of biological markers in epidemiological research. For some situations, suitable markers do not exist or the properties of available markers are inadequate to meet the study's objectives. For example, analysis of the marker may be too expensive, or available markers may not be specific enough for a particular exposure. In other cases, the properties of a potentially useful marker may not be known. Answers may yet be needed to such questions as "What is the relationship between levels of the marker in the surrogate tissue and levels of the marker in the target tissue?"; "What are the pharmacokinetic properties of the marker?"; or "How stable is the marker when stored frozen?"

In addition, markers may not have been validated for the applications intended by the researcher. Validation of biological markers is a complicated issue. Methods of validating a marker differ in some respects, depending on whether the marker is used to measure an exposure, an intermediate end point, or a disease outcome. In all three cases, steps that should be undertaken include preliminary laboratory studies to establish the limits of detection and reliability of the assay. Preliminary field studies in humans should be undertaken (1) to establish that the assay can detect population levels of the marker; (2) to determine the range of marker levels in human populations; (3) to identify the factors associated with variation in marker levels, such as race, gender, and age; and (4) to determine the reliability of the assay under study conditions (35). All of this information is known for only few markers.

In the case of exposure markers and markers of intermediate end points, preliminary laboratory work should also include animal experiments to describe dose–response relationships between known doses of an external exposure and internal marker levels. Information should be available for markers measured after acute and chronic exposure situations (36). Human field studies could then be conducted to assess the dose–response relationships between known levels of an external exposure and internal marker levels in high- versus low-exposure situations. One would want to see that, on average, those with the high exposures are more likely to have higher levels of the marker than those with low exposures. Finding a perfect correlation between levels of an external exposure and marker levels does not necessarily invalidate the marker. In fact, a strength of biological markers is that they integrate exposures from many sources and provide estimates of exposure to the individual that consider the use of protective equipment and interindividual differences in, for example, absorption, metabolism, and excretion. Thus, even if the marker does not perfectly correlate with external measures of the exposure, it may accurately indicate internal exposure to the individual. External exposure information, such as can be obtained by area air monitoring and questionnaires, needs to be assessed to characterize the contribution of other sources of exposure to marker levels. Data concerning the pharmacokinetic properties of the marker would be needed to interpret marker levels.

More research has concentrated on establishing associations between external levels of exposures and internal marker levels than on determining the predictive value of the marker levels. Ultimately, it would be helpful to show the link between elevated marker levels and increased risk of some disease outcome as an indicator of the validity of the marker.

Markers of disease can be validated against a medically determined "gold standard" for diagnosing the disease. As part of the validation process, one would calculate the sensitivity, specificity, and positive predictive value of the marker (37). Sensitivity, as defined by epidemiologists is the probability that the marker will correctly identify an individual as having the disease, given that the individual actually has the disease (as determined by the gold standard method). The specificity is the probability that the marker will correctly identify the individual as not having the disease, given that the individual actually does not have the disease. The positive predictive value is the probability that the individual has the disease, given that the marker indicates the presence of the disease. In each case, one would want these values to be as high as possible. For example, a recent study (38) evaluated elevated levels of the angiogenic peptide basic fibroblast growth factor in urine as a marker for bladder cancer among bladder cancer patients diagnosed with cytoscopic analysis and radiological studies and found the sensitivity and specificity to be modest at 81% and 64%, respectively. If one is trying to validate a marker of preclinical disease, some follow-up of study participants will be necessary to determine if the individuals who test positive for the marker actually develop the disease.

Quality-Control Issues

To produce meaningful study results, certain quality-control procedures should be followed. Laboratory personnel should be blinded to the exposure and disease characteristics of the participants from whom the biological specimens were obtained. Specimens from exposed–nonexposed or disease–nondiseased individuals should simultaneously be analyzed to avoid confusion as to whether differences in the characteristic being studied or differences in laboratory techniques over time are responsible for any observed differences in laboratory results between groups. In addition to analyzing specimens from each of the study participants, it is recommended that one analyze specimens known to be positive and those known to be negative for the marker of interest. Also, analyzing more than one specimen from the same individual for a subset of the study participants aids in assessing the reliability of the assay.

Collaborative Research Involving Biological Markers

Epidemiological research, in general, is becoming more and more collaborative. People in vastly different fields are working together. This is especially true in biological marker research. These collaborations bring with them benefits and challenges. The major benefit is that questions can be addressed by collaborative efforts that are less well addressed by the members of the individual fields working separately. The interactions between the collaborators from different fields can lead to creative solutions to problems.

The challenges inherent in collaborative work are many. Problems arise because collaborators from different fields think differently, use the same words to refer to different things, and have different agendas (23, 33). Thus, the very reasons why investigators form collaborations present obstacles to the collaborative effort. For collaborative efforts to succeed, members of each discipline must learn the language, strengths, and weaknesses of the other discipline.

Epidemiologists and laboratory scientists differ in their outlook on a number of issues. Whereas laboratory scientists are often concerned about the effects of agents on individuals or small groups of subjects, epidemiologists are concerned with making inferences about disease causation in populations. Epidemiologists can say little about the definite cause of a disease in a specific individual. They can only estimate the probability that a particular agent led to the development of the disease in that individual, on the basis of population statistics.

The time frame of laboratory research tends to be short, compared with that of epidemiological research. Laboratory experiments can be completed in days to months, whereas epidemiological research often takes years to complete. In the Framingham Heart Study (39), for example, participants have been followed for over 20 years. Even in studies in which participants are not followed over time, it can take months to years to identify eligible study participants, enroll them in the study, and obtain specimens from them.

The sample sizes in laboratory studies tend to be small, compared with those of epidemiological studies. Laboratory research usually involves a handful of animals exposed to different doses of a chemical. Epidemiological research often must involve hundreds or thousands of participants to produce statistically significant results. Laboratory scientists may not be equipped either to handle the large number of specimens required of many epidemiological investigations or want to commit resources for such a long time.

The number of extraneous factors in laboratory experiments also tends to be small, compared with those in epidemiological research. Laboratory researchers often work with genetically similar animals and control as many other factors as possible. Epidemiologists observe what happens in the real world.

Participants in epidemiological studies are often not genetically similar and have a variety of different exposures and life style habits. Epidemiologists try to control for as many factors as possible in the statistical analysis of the data.

Epidemiologists and laboratory scientists often have different views of the laboratory assays themselves. Laboratory scientists tend to want to perfect the assay, for example, by making it more sensitive. Once the assay is "perfected," the laboratory scientist may no longer be interested in it. The epidemiologists, on the other hand, want to use the same assay over and over again on all the participants in the study.

The differences between the two disciplines can be either strengths or obstacles. If laboratory scientists and epidemiologists work together, they can learn to maximize the strengths and minimize the obstacles. For example, epidemiologists and laboratory scientists may need to develop more clever and specific hypotheses so that large numbers of subjects are not necessary to obtain meaningful study results. Because the operation of laboratory assays in population settings is often different from that in the controlled experimental environment (A. Wilcox: personal communication), technical aspects of the assay often need to be addressed. Collaborations between laboratory scientists and epidemiologists have the potential to provide interesting avenues of research for investigators from both disciplines.

Examples of the Use of Biological Markers in Environmental and Occupational Epidemiology

Exposure Assessment. Among the most difficult aspects of determining the health consequences of environmental exposures is identifying who is exposed and quantifying the levels of exposure of the affected individuals. Biological markers of exposure may help alleviate some of these problems. An example of a situation in which biological markers may be useful is trying to assess the possible adverse health effects of living near a hazardous waste site. Once it is determined that the agents contained in the site have the potential to cause adverse human health effects (for example, through literature reviews), the next step is to determine whether the residents near the site are actually exposed to the agents in the hazardous waste. The ideal biological marker of exposure, in this case, would be specific for a particular agent present at the waste site. If one used a marker that was not very specific, confusion might arise as to whether persons positive for the marker were exposed because of agents emanating from the waste site or because of other exposures. When less specific markers are used, sources of other exposures can sometimes be assessed by questionnaires. The Agency for Toxic Substances and Disease Registry (ATSDR) (40) assessed levels of arsenic in the urine of residents living near an abandoned arsenic production site. Urinary arsenic levels (μg/g creatinine) were not statistically significantly higher among all potentially exposed residents of the nearby apartment complex, compared with controls who were less likely to have been exposed. However, three children in the nearby apartment complex had particularly high urinary arsenic levels. The three children and two adults with arsenic levels greater than 50 μg/g creatinine were retested after a warning to stay away from the contaminated area. The lowered arsenic levels among the retested children suggest that the site was the source of the arsenic exposure and that the warning may have been effective in reducing exposure.

Intermediate End Points. Most of the environmental research involving biological markers has proceeded by identifying an exposure and then determining its effect on the levels of a particular marker. This is exemplified in occupational settings in which the blood cells of workers exposed to chemicals, such as ethylene oxide and petroleum vapors, have been evaluated for the presence of various markers of biological response, including chromosome aberrations, sister chromatid exchanges, and micronuclei (25,41). In many cases, elevations in these intermediate markers have been associated with the exposure of interest. Although evidence of increased chromosomal breakage is generally not considered favorable, no one knows what having elevated levels of these markers really means (42). Here is an example of a situation in which research is needed to determine the predictive value of markers for adverse health outcomes. Knowledge of their predictive value could be beneficial for understanding disease causation and preventing disease at earlier stages.

Summary

Epidemiology is the study of the distribution and determinants of diseases in populations. Traditionally, epidemiologists have taken a "black box" approach to the study of environmentally induced disease, assessing an environmental or external exposure and measuring some health outcome. It is hoped that, through the use of biological markers, often assessed by molecular and toxicological techniques, we can begin to understand the intermediate steps between exposure and disease occurrence and can improve the precision of exposure and outcome measurements. A biological marker is a cellular, biochemical, or molecular alteration that is measurable in biological media such as human tissues cells or fluids. Biological markers can be classified into categories that represent a sequence of events from exposure to disease, including markers of internal dose, biologically effective dose, early response, altered structure and function, and disease. Outside this sequence of events are susceptibility factors that can influence events at any point along the pathway. Because epidemiological investigations often include large numbers of participants, the biological markers that are most desirable for use in human studies are those that can be measured with a minimum amount of skill or equipment in a small amount of biological medium and can be obtained with minimally invasive techniques. The assays used to detect the markers should be inexpensive, sensitive, specific, reliable, and able to quantify accurately levels of the markers present in human specimens. Unfortunately, the properties of many potentially useful markers are still unknown, and the properties of available markers may not be appropriate for particular applications. Future research in this area will require the collaborative efforts of epidemiologists and laboratory scientists, among other professionals. Collaborative efforts can lead to innovative research. For the collaborations to succeed, researchers from each of the various disciplines must learn the language, strengths, and weaknesses of the other disciplines.

References

1. Hulka, B. S. In *Biological Markers in Epidemiology*; Hulka, B. S.; Wilcosky, T. C.; Griffith, J. D., Eds.; Oxford University: New York, 1990; Chapter 1, pp 3–15.

2. Wilcosky, T. C. In *Biological Markers in Epidemiology;* Hulka, B. S.; Wilcosky, T. C.; Griffith, J. D., Eds.; Oxford University: New York, 1990; Chapter 3, pp 28–55.
3. Hulka, B. S.; Wilcosky, T. C. *Arch. Environ. Health* **1988**, *43(2)*, 83–89.
4. National Research Council. *Biologic Markers in Reproductive Toxicology;* National Academy: Washington, DC, 1989; p 17.
5. Perera, F. P.; Weinstein, I. B. *J Chron. Dis.* **1982**, *35*, 581–600.
6. Mushak, P. In *Lead Exposure and Child Development and International Assessment;* Smith, M. A.; Grant, L. D.; Sors, A. I., Eds.; Kluwer Academic: Dordrecht, The Netherlands, 1989; Chapter 22, pp 129–145.
7. Langone, J. J.; Gjika, H.; Van Vunakis, H. *Biochemistry* **1973**, *12*, 5025–5030.
8. Swaen, G. M.; Slangen, J. J.; Volovics, A.; Hayes, R. B.; Scheffers, T.; Sturmans, F. *Br. J. Ind. Med.* **1991**, *48(2)*, 130–135.
9. Goldring, J. M.; Lucier, G. W. In *Biological Markers in Epidemiology;* Hulka, B. S.; Wilcosky, T. C.; Griffith, J. D., Eds.; Oxford University: New York, 1990; Chapter 5, pp 78–104.
10. Segerback, D. *Chem. Biol. Interact.* **1983**, *45*, 139–151.
11. Wogan, G. N.; Gorelick, N. J. *Environ. Health Perspect.* **1985**, *62*, 5–18.
12. Murphy, S. E.; Palomino, A.; Hecht, S. S.; Hoffmann, D. *Cancer Res.* **1990**, *50*, 5446–5452.
13. Perera, F. P.; Santella, R. M.; Brenner, D.; Poirier, M. C.; Munshi, A. A.; Fischman, H. K.; Van Ryzin, J. *J. Natl. Cancer Inst.* **1987**, *79(3)*, 449–456.
14. Stich, H. F.; Rosin, M. P. In *Carcinogens and Mutagens in the Environment.* Stich, H. F., Ed.; CRC: Boca Raton, FL, 1983; Vol. 2, Chapter 2, pp 17–25.
15. Schmid, W. *Mutat. Res.* **1975**, *31*, 9–15.
16. Jenssen, D. In *Sister Chromatid Exchange;* Sandberg, A. A., Ed.; Alan R. Liss: New York, 1982; Chapter 4, pp 47–63.
17. Vine, M. F.; McFarland, L. T. In *Biological Markers in Epidemiology;* Hulka, B. S.; Wilcosky, T. C.; Griffith, J. D., Eds.; Oxford University: New York, 1990; Chapter 10, pp 196–213.
18. Cleaver, J. E. *Nature (London)* **1968**, *218*, 652–656.
19. Ayesh, R.; Idle, J. R.; Ritchie, J. C.; Crothers, M. J.; Hetzel, M. R. *Nature (London)* **1984**, *312*, 169–170.
20. Cartwright, R. A.; Glashan, R. W.; Rogers, H. J.; Ahmad, R. A.; Barnham-Hall, D.; Higgins, E.; Kahn, M. A. *Lancet* **1982**, *ii*, 842–846.
21. Schmidt, W.; Popham, R. E. *Cancer* **1981**, *47*, 1031–1041.
22. Ross, R. K.; Yuan, J. M.; Yu, M. C.; Wogan, G. N.; Qian, G. S.; Tu, J. T.; Groopman, J. D.; Gao, Y. T.; Henderson, B. E. *Lancet* **1992**, *339*, 943–946.
23. Hulka, B. S. *Cancer Epidemiol. Biomed. Prev.* **1991**, *1*, 13–19.
24. Taylor, J. A.; Sandler, D. P.; Bloomfield, C. D.; Shore, D. L.; Ball, E. D.; Neubauer, A.; McIntyre, O. R.; Liu, E. *J. Natl. Cancer Inst.* **1992**, *84*, 1626–1632.
25. Stolley, P. D.; Soper, K. A.; Galloway, S. M.; Nichols, W. W.; Norman, S. A.; Wolman, S. R. *Mutat. Res.* **1984**, *129*, 89–102.
26. Dar, E.; Kanarek, M. S.; Anderson, H. A.; Sonzogni, W. C. *Environ. Res.* **1992**, *59*, 189–210.
27. Greenland, S.; Robins, J. M. *Am. J. Epidemiol.* **1985**, *122*, 495–506.
28. Stich, H. F.; San, R. H.; Rosin, M. P. *Ann. N.Y. Acad. Sci.* **1983**, 93–105.
29. Jarvis, M. J.; Russel, M. A. H.; Benowitz, N. L.; Feyerabend, C. *Am. J. Publ. Health* **1988**, *78(6)*, 696–698.
30. Benowitz, N. L.; Kuyt, F.; Jacob, P., III. *Clin. Pharmacol. Ther.* **1982**, *32*, 758–764.

31. Skipper, P. L.; Tannenbaum, S. R. *Carcinogenesis* **1990,** *11(4),* 507–518.
32. Bartsch, H.; Caporaso, N.; Coda, M.; Kadlubar, F.; Malaveille; Skipper, P. L.; Talaska, G.; Tannenbaum, S. R.; Vineis, P. *J. Natl. Cancer Inst.* **1990,** *82,* 1826–1831.
33. Hulka, B. S.; Margolin, B. H. Methodological Issues in Epidemiologic Studies Using Biologic Markers. *Am. J. Epidemiol.* **1992,** *135(2),* 200–209.
34. Wilcosky, T. C.; Griffith, J. D. In *Biological Markers in Epidemiology;* Hulka, B. S.; Wilcosky, T. C.; Griffith, J. D., Eds.; Oxford University: New York, 1990; Chapter 2, pp 16–27.
35. National Research Council (U.S.)—Board on Environmental Studies and Toxicology; Committee on Advances in Assessing Human Exposure to Airborne Pollutants; Commission on Geosciences, Environment and Resources. In *Human Exposure Assessment for Airborne Pollutants: Advances and Opportunities;* National Academy of Science: Washington, DC, 1991; Chapter 4, pp 115–142.
36. National Research Council. *Biological Markers in Pulmonary Toxicology;* National Academy of Science: Washington, DC, 1989; pp 11–42.
37. Schulte, P. A. *Am. J. Epidemiol.* **1987,** *126,* 1006–1016.
38. Nguyen, M.; Watanabe, H.; Budson, A. E.; Richie, J. P.; Folkman, J. *J. Natl. Cancer Res.* **1993,** *85(3),* 241–242.
39. Fraser, G. E. *Preventive Cardiology.* Oxford University: New York, 1986; p 43.
40. *Final Report of Technical Assistance to the Houston Health and Human Services Department—Crystal Chemical Company Arsenic Exposure Study;* Agency for Toxic Substances and Disease Registry (ATSDR), Division of Health Studies, Division of Health and Human Services: Atlanta, GA, July 1989.
41. Hogstedt, B.; Gullberg, B.; Mark-Vendel, E.; Mitelman, F.; Skerfving, S. *Hereditas* **1981,** *94,* 179–187.
42. Schwartz, G. In *Biological Markers in Epidemiology;* Hulka, B. S.; Wilcosky, T. C.; Griffith, J. D., Eds.; Oxford University: New York, 1990; Chapter 8, pp 147–172.

RECEIVED for review September 3, 1992. ACCEPTED revised manuscript April 8, 1993.

10

Examples of Measuring Internal Dose for Assessing Exposure in Epidemiological Studies

Larry L. Needham

Division of Environmental Health Laboratory Sciences, National Center for Environmental Health, Centers for Disease Control and Prevention, U.S. Department of Health and Human Services, Atlanta, GA 30341–3724

> *Exposure to environmental contaminants occurs when the contaminant is present in the environment and humans have contact with that environment. Environmental public health scientists, especially epidemiologists, study the relationship between such exposure and any adverse health effects. The exposure is often assessed on the basis of measured concentrations of the contaminants in the environment and the estimated duration of human exposure. However, frequently these models are not predictive of the amount of toxicant absorbed in the human; thus, there is a need to measure the internal dose of these toxicants in humans. This measurement, along with pharmacokinetic information for the toxicant, is the best marker of exposure with which to relate adverse health effects. Several examples that illustrate the need for measuring the internal dose are presented.*

EPIDEMIOLOGISTS AND OTHER ENVIRONMENTAL public health scientists are interested in the relationship between exposure assessment and adverse health effects or at least a biochemical change that may be a marker of a potential adverse health effect. To study this relationship, they may use three basic types of observational analytical investigations: the case-control study; the cohort study, prospective and retrospective; and the cross-sectional study. In a "case-control study", a case group that has the adverse health effect and a control, or comparison, group that does not have the effect are selected for investigation; the proportions with the exposure of interest in each group are compared. Thus, in a case-control study, the start-

This chapter not subject to U.S. copyright
Published 1994 American Chemical Society

ing point (the adverse health effect that has occurred prior to the initiation of the study) is the adverse health effect. Exposures to several pollutants that may lead to the adverse effect can be studied. In a "cohort study", however, participants are identified on the basis of their exposure status. In a prospective cohort study, the adverse health effect has not occurred at the time of the initiation of the study; in a retrospective cohort study, the investigation is initiated after the exposure and effect have occurred. Several health effects that might result from exposure to the pollutant can be studied. Finally, in the "cross-sectional survey", the status of an individual with respect to the presence or absence of both exposure and disease is assessed simultaneously. However, in all of these studies, exposure and health effects must be assessed. In environmental public health, the exposure is frequently to ubiquitous environmental pollutants, including dioxins, polychlorinated biphenyls, and volatile organic solvents. Therefore, because most all people have been in contact with these pollutants, exposure is not a qualitative yes-or-no decision; ideally, it is quantified, but at least it must be "scientifically estimated." What is the best means of doing this?

Figure 1 illustrates the exposure and health effects pathway (1,2), which we have amended (3). As shown in Figure 1, a pollutant originates from a source and is emitted into the environment, where it can undergo chemical and physical interactions and where it may accumulate. Humans are exposed when, for an interval of time, they come into contact with that pollutant in an environmental medium—soil, water, air, or food. The concentration of a given pollutant in the environmental media with which humans have contact, integrated over the time of contact, is called the external dose (3) or

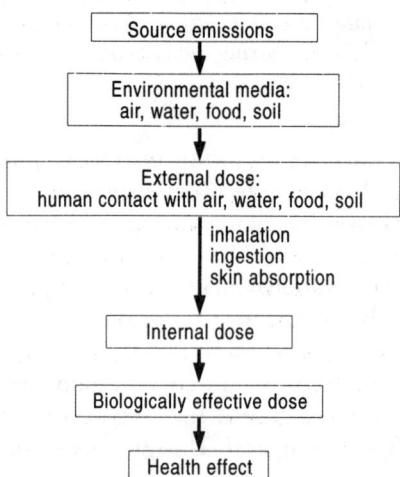

Figure 1. *Pathway of a contaminant from emission to resulting health effect.*

potential dose (2). When humans have contact with these environmental media, the pollutants enter the body via inhalation, ingestion, or skin absorption. Once in the body, a pollutant may be distributed throughout it.

The amount of contaminant absorbed in body tissue is called the internal dose. Common measures of internal dose are the blood and urine levels of pollutants or their metabolites. A portion of this internal dose may reach and interact with a target site over a given period so as to alter physiological function; this portion is called the biologically effective dose.

Thus, at least three measures or estimates of dose may be used in assessing human exposure. These are external dose, biologically effective dose, and internal dose. Each has certain advantages and disadvantages.

External Dose

Traditionally, environmental epidemiologists have assessed exposure by estimating the external dose, that is, the total concentration of a given toxicant in the environmental media with which humans come into contact, integrated over the time of contact. The concentration of the pollutant in the environmental media is sometimes based on analytical measurements of environmental samples—water, air, soil, and food—collected at the exposure site near the time of exposure or as close to it as possible. Perhaps the "best" environmental sample of an airborne pollutant would be an air sample collected at the time of exposure by a personal air monitor located in the breathing zone. Clearly, this is never achieved, except in designed experiments. The time of contact with the environmental media containing the pollutant is generally collected by questionnaire and history. This combination of questionnaire and history and environmental measurements are then weighted into an exposure index, which is used as an estimate of exposure for each person.

Exposure indices have several potential problems, particularly bias of administering and answering the questionnaire. This bias may arise whenever noncomparable information is obtained from the different study groups. This may result from the interviewer eliciting or interpreting the information in different ways (interviewer bias); it may also result from the participants intentionally or unintentionally reporting the events in noncomparable manners (recall bias), for example, having problems recalling the frequency of walking on contaminated soil or consumption of a certain food. Another source of error is the inability to adjust for individual factors related to how much toxicant enters the body and how much is absorbed (individual metabolism differences, individual nutritional status during exposure, and personal habits such as hand-to-mouth activities). Although generally fewer problems are associated with the quantitative measurements in the environmental media, these measurements may not be representative of the degree of contamination at the time the human exposure occurred; for example, the average level of a pollutant in fish caught in a river may not be representative of all such fish in that river. Thus, we believe that estimates of the external dose are not the best means of assessing exposure.

Biologically Effective Dose

By definition, the best measure of exposure for assessing dose–response relationships is the "biologically effective dose". Ideally, environmental health scientists would like to have sensitive and specific measurements of the biologically effective dose. However, identifying the target site of the toxicant is a major impediment to using measures of the biologically effective dose to quantify exposure, and even when the target site is known, a highly invasive procedure may be required to sample that site (e.g., liver or brain). Some organic toxicants, or their metabolites, covalently bond to DNA, thus forming a DNA adduct. If this adduct is at a site that leads to an altered reading of the genetic code, it may lead to mutagenicity and cancer in humans. However, other adducts that lead to no such alteration may form. Because we cannot analytically distinguish between these adducts, the analyses of human samples, such as circulating lymphocytes, for all such adducts are generally considered measurements of the biologically effective dose. Likewise, measurements of adducts with hemoglobin and proteins, such as albumin, have also been considered measurements of the biologically effective dose (4). Some of these adducts are specific markers for toxicants (e.g., benzo(a)pyrene in lymphocytes), whereas others are much less specific (e.g., DNA adducts with alkyl groups). The measurement of adducts in humans is still in the experimental development stage, and much more information is needed before it can be used as a quantitative measurement of exposure, although it can be used as a marker of exposure. Other disadvantages of these measurements are that sample throughput is somewhat problematic for moderate-size epidemiological studies, and very sensitive analytical methods are required.

Internal Dose

The next most useful exposure measures are those of internal dose. The direct measurement of a toxicant or one of its metabolites in blood or urine has significantly improved human exposure assessment and has thus improved human risk assessment for a number of important toxicants. For example, without blood lead measurements, most of the central nervous system effects of low-level lead exposure could not have been detected.

To interpret blood or urine toxicant levels accurately, analysts must know the pharmacokinetics of the toxicant and also must know the background levels found in the general population. For example, some toxicants, such as the volatile organic compounds, are rapidly eliminated, whereas others, such as the chlorinated hydrocarbon pesticides, may have a half-life in humans of more than 5 years. Thus, this information is critical for interpreting whether the measured concentration reflects recent exposures or long-term (chronic) exposures, or both. It is still essential that the epidemiologist collect, to the extent possible, nonbiased questionnaire and history data on the potential exposure.

If the compound can be detected in blood or its components, we prefer these matrices because (1) the concentration of the toxicant in the urine can vary greatly throughout the day; (2) the level in the blood more likely reflects the concentration available to the target site; and (3) although the parent compound is normally measured in blood, the metabolite is usually measured in urine, and that metabolite may not be specific for only one toxicant. A potential disadvantage for the measurement of the internal dose in blood is that the concentration of lipophilic compounds in an individual can vary throughout the day, depending on the variability in the lipid content of the serum. However, expressing the concentration of the toxicant on a lipid-weight basis, instead of a whole-weight basis, has proved to be an adequate adjustment (5).

In occupational studies, on the other hand, urine is the preferred biological matrix, primarily because it can be procured noninvasively. Its major advantage is the ease in obtaining the sample compared with the difficulty of venipuncture for obtaining a serum sample. The American Conference of Governmental Industrial Hygienists (6) publishes biological exposure indices (BEIs) for use in assessing occupational exposure to selected chemicals. The BEI is a reference value that indicates typical concentrations of selected chemicals or their metabolites, primarily in urine but also in blood or exhaled air, for healthy workers who have been exposed primarily by inhalation to the selected chemical at its threshold limit values. It is generally believed that there is poor correlation between the levels of contaminants in the blood and urine; however, we found a high correlation of the levels of pentachlorophenol in these two matrices after the urine was corrected for creatinine (7).

We recognized the critical need for sensitive and specific measurements of internal dose in order to improve human exposure assessment. Our laboratory is committed to developing such measurements and to applying them to epidemiological studies, which should substantially improve human risk assessment. We have developed and applied methods of measuring the internal dose of polychlorinated and polybrominated biphenyls in serum (8) as part of the Michigan Polychlorinated Biphenyl–Polybrominated Biphenyl Study (9); pentachlorophenol (10) in the urine and serum of residents of log houses (7); low-molecular-weight diols in serum (11) of alcoholics (12); phenol and cresols in the urine of workers (13); dioxin in the adipose tissue (14) of residents of Missouri (15–18); dioxin in the serum (19) of residents of Seveso, Italy (20); phenoxy acid herbicides and chlorinated phenols in urine (21) of residents living near a Superfund site (22); and the volatile organic compounds in the blood of U.S. residents (23). We give other examples of internal-dose measurements further on.

Sometimes in an epidemiological study, an exposure index based on measurements in environmental media and on questionnaire and history information is all that is practicable and must be used. Whenever possible, this index should be compared with measures of internal dose on at least a subset of the exposed population to validate the exposure index and to calibrate the levels of exposure being examined. This was effectively done in studying worker exposure to dioxin (24). However, measurement of the internal dose is not the panacea for assessing exposure. As pointed out by Hulka and Margolin (25), variability within the laboratory and also biological variability, both intraperson and interperson, can lead to errors in the assessment of exposure. The laboratory personnel and epidemiologists must attempt to minimize these errors by using, for example, proper sampling design and quality assurance procedures.

In this chapter we present examples of our experiences in case-control and cohort epidemiological studies in which we could compare measurements of internal dose with exposure indices. The situations range from occupational exposure to exposure among residents living near hazardous waste sites.

Examples

Operation Ranch Hand Study. From 1962 to 1970, during the Vietnam conflict, the main mission of the U.S. Air Force's Operation Ranch Hand was to spray defoliants, such as Agent Orange, over densely vegetated areas of South Vietnam. Agent Orange consisted of an equal mixture of (2,4-dichlorophenoxy)acetic acid (2,4,-D) and (2,4,5-trichlorophenoxy)acetic acid (2,4,5-T) in diesel oil; the 2,4,5-T was contaminated with 2,3,7,8-tetrachlorodibenzo[b,e][1,4] dioxin (2,3,7,8-TCDD [dioxin]) in the parts-per-million range. Dioxin is lipid-soluble and thus tends to store in the lipid-rich depots of the human body; it has a long half-life of more than 7 years in humans (26). Therefore, quite a bit is known about its pharmacokinetics in humans. In 1982, the Air Force designed a prospective cohort study, specifically to study health outcomes that may be associated with exposure to Agent Orange and other herbicides containing dioxin. These health studies will be performed on the veterans of Operation Ranch Hand every 5 years through the year 2002. One of the first tasks was to develop an exposure index, which is the concentration of the contaminant multiplied by the time of contact, to be the basis of assessing human exposure.

This exposure situation was similar to that of an occupational setting in that the primary exposure was thought to be direct exposure to the herbicide itself, rather than indirect exposure through an environmental pathway. The Air Force knew the average concentration of dioxin in the Agent Orange during various periods. The duration and frequency of potential exposure for the men in different military occupational groups within the operation were estimated on the basis of military records. The number of men in the study was limited to the 1200–1300 survivors of the operation. The U.S. Air Force and various review boards believed not only that this exposure index could serve as a reliable basis for assessing exposure to dioxin, but that any noted adverse health effects could be related to this index.

In 1987, the U.S. Air Force contracted with our laboratory to analyze 150 serum samples from Operation Ranch Hand veterans in order to compare the Air Force's exposure index with the measured internal dose of the veterans. There was essentially no correlation between the exposure index and the serum dioxin level. Therefore, the Air Force then contracted with the Centers for Disease Control and Prevention (CDC) to analyze the serum of all surviving members of Operation Ranch Hand. This lack of correlation between the internal dose and the Air Force's exposure index continued throughout the study (27). There may be several reasons for this lack of correlation; these include the Air Force not knowing which bases in Vietnam housed the Agent Orange during a given time period and individual differences in job performance and other exposure activities, such as using Agent Orange as a degreaser for oil on the body. The Air Force subsequently changed the basis of its exposure assessment to the serum dioxin measurement, a measure of internal dose. Had the Air Force used its original exposure index for the Operation Ranch Hand study, any conclusions about health effects would have been invalid.

U.S. Army Ground Troops in Vietnam. The contaminant of concern was dioxin in Agent Orange; the potential environmental pathways were skin contact with and inhalation of the spray containing the herbicide, skin contact with sprayed vegetation and soil, and ingestion of water and food that had been sprayed. The amount of dioxin in the Agent Orange from 1966 to 1969 was known. The duration of contact was gathered from questionnaires given to the veterans and from U.S. military records of the locations of military units, areas where herbicide was sprayed, and the dates when herbicide was sprayed.

Six exposure indices were generated from this information; four of the indices were based on a soldier's potential for exposure from direct spray or on his being in an area that had been sprayed within the previous 6 days; the other two exposure indices used self-reported data and included an index based on the veteran's perception of how much he was exposed. To test the validity of these exposure indices, we measured blood dioxin levels in a sample of enlisted ground troop veterans who had served in III Corps for an average of 300 days during 1966–1969. For comparison, dioxin in a sample of non-Vietnam U.S. Army veterans who served during the same time was also measured (28).

The results showed no meaningful association between dioxin levels and any of the exposure indices. The mean, median, and frequency distributions for both the Vietnam and non-Vietnam veterans were remarkably similar, indicating little, if any, increased exposure to dioxin in this population (Table I). This example illustrates the value of measurements of internal dose in exposure assessment. It also shows the need to develop specific and sensitive methods: If the detection limit for dioxin had been 20 ppt (lipid-adjusted), almost all the results would have been "nondetect." Furthermore, because elevated exposures could not be documented, plans for a prospective cohort health study were dropped.

Seveso, Italy. More than 15 years have passed since the release of dioxin-containing materials from the Icmesa chemical plant near Seveso, Italy. This incident has received perhaps more public attention and study than any other residential exposure to a synthetic environmental toxicant; yet no deaths have been directly attributed to it. In fact, in its last meeting of February 19–21, 1984, the International Steering Committee of experts noted that chloracne was the only clinical alteration to have a positive association with soil levels of dioxin. The committee also reported that no conclusion was yet possible about its association with cancer, the study of which is still ongoing. Following the 1976 to 1985 medical examinations of the Seveso people and a nearby comparison group, small amounts of unused serum were frozen. In 1988, we met Italian officials and outlined a protocol aimed at first determining whether we could measure dioxin in the small amount of reserved serum available from the most exposed population (based on residential soil levels of dioxin); if so, could we determine whether there is a difference between the dioxin levels in those who had developed chloracne (10 people) and those who had not (10 people). Many factors were similar for the two groups (e.g., gender and dioxin soil levels), whereas other factors favored higher exposure of the nonchloracne group (e.g., greater tendency to eat local produce and poultry, average number of days residing in the most exposed zone following the explosion, and average length of time from explosion to the blood draw); these characteristics are summarized in Table II. The other primary difference is age at the time of the explosion; for chloracne cases, the range was 2–16 years with a mean of 7.2 years, and for comparisons it was 15–71 years with a mean of 42.6 years.

Individuals who had the highest lipid-adjusted serum TCDD levels tended to develop chloracne. However, there was no clear serum TCDD level above which chloracne was produced and below which chloracne did not occur (20); such a result would have been surprising. The point is that there was no "magic" level that produced chloracne. The best quantitative marker for exposure is not chloracne or residential TCDD soil levels but rather the internal dose.

Table I. Distribution (1987) of 2,3,7,8-Tetrachlorodibenzo-p-dioxin in Serum Lipids (PPT) of 743 Vietnam-Era Veterans

Place of Service	Median	Mean ± SD	Percentile				
			25th	50th	75th	90th	95th
Non-Vietnam ($n = 97$)	3.8	4.1 ± 2.3	3.8	3.8	4.9	7.2	9.2
Vietnam ($n = 646$)	3.8	4.2 ± 2.6	2.8	3.8	5.1	6.8	7.8

Residents Living near Toxic Waste Sites. Work practices that involved electric transformers with polychlorinated biphenyl (PCB)-containing dielectric fluid at a repair and maintenance facility in Paoli, Pennsylvania, resulted in extensive, high-level PCB contamination of soil and surfaces—up to 420,000 ppm in the work area and 36,000 ppm in soil. Soil contamination in nearby residential neighborhoods ranged from less than 1 ppm to more than 6400 ppm; the normal level of PCBs in U.S. soil is less than 1 ppb. Local creeks and trout in those creeks also contained elevated levels of PCBs. In this example, the frequency and duration of contact with the contaminant were not estimated for each person or for groups. Rather, 62 persons older than 3 years of age, from neighborhoods adjacent to or immediately down a gradient from the repair facility were selected by using a probability sampling to represent the entire population of the neighborhoods (396 persons).

A second preselected group of 85 was chosen from persons who lived in homes with yards that had at least one soil PCB measurement greater than 150 ppm. We measured the PCBs in the serum (29) of these 85 persons. In the probability sampling group, the serum PCB levels ranged from less than 1 ppb to 23.7 ppb with a geometric mean of 4.3 ppb and an arithmetic mean of 5.8 ppb. The geometric mean and arithmetic mean of the 23 serum PCB concentrations in the preselected group were 5.9 ppb and 10.4 ppb, respectively—higher than the probability sampling group—but the geometric means were not statistically significantly different. If the one outlier value of 76.9 ppb is discarded, the geometric and arithmetic means are 5.3 ppb

Table II. Characteristics (Mean ± SD) of Selected Seveso A Zone Residents

Group	N	Age (Years)	Dioxin Soil Levels ($\mu g/m^2$)	Days in A Zone after Explosion	# Who Ate Local Produce– Poultry	Dioxin Serum Levels—Lipid Adjusted (ppt)
Chloracne	10	7.2 ± 4.2	1071 ± 338	10 ± 7.0	6	19,144 ± 16,241
Nonchloracne	9	42.2 ± 16.1	1171 ± 26	16.1 ± 0.4	9	5240 ± 2946

NOTES: SD is standard deviation; ppt is parts per trillion

and 7.4 ppb, respectively. This value was discarded because the person with this value had been occupationally exposed; PCB values for other members of his family were between 5 and 9 ppb, and the next highest PCB level found was less than half of this level.

The authors concluded that the population near the work site in Paoli had exposure similar to that of populations with typical background exposure elsewhere in the United States. Multiple linear regression analysis, controlled for age, sex, years of residence near site, and other factors, found age to be the only variable significantly associated with serum concentrations of PCBs (30), which, like dioxin, are lipophilic and have long biological half-lives (31). One unusual finding in this study was that gas chromatographic peaks consistent with a rarely used Aroclor (AR 1268) were found in three persons and in 9 of 16 dogs (32). However, environmental testing did not reveal the presence of AR 1268.

These PCB findings are similar to other site-specific investigations that we have conducted around toxic waste sites. In 1988, Stehr-Green et al. (33) reported that in 10 of 12 pilot exposure-assessment studies, we found no excess proportion of overtly exposed persons, in spite of high PCB levels in soil or leachate on the sites. The two sites where an elevated proportion of exposed persons was found were in Newport County (New Bedford), Massachusetts, and Monroe County (Bloomington), Indiana. A follow-up study of the population around Bloomington, Indiana, has been completed, and the report is in draft form. The Greater New Bedford Study (34) was conducted from 1984 to 1987 to assess the prevalence of elevated levels of PCBs in the serum of persons aged 18 to 64 years who had resided in the area for at least 5 years. Because of documented environmental contamination by PCBs in the New Bedford area, the practice of recreational fishing in the harbor for food, and the findings from a pilot study (35), researchers expected to find a significant number of persons with elevated PCB levels. Instead, they found that the PCB levels among the 840 participants were similar to those found among persons in other urban U.S. populations; the geometric means of the samples were 4.3 ppb among males and 4.2 ppb among females. However, persons who ate locally caught seafood from inner New Bedford Harbor tended to have higher concentrations of serum PCBs, although their mean value was less than 20 ppb, which is the upper 95% level of serum PCBs in the United States (32).

We also collaborated in an exposure assessment and health effect study of people living near the Hollywood Dump Site, in Memphis, Tennessee; this toxic waste site contains chlorinated hydrocarbon pesticides. The results of the analyses (36) of the serum of 370 persons and the adipose tissue of 297 of them revealed that persons living around the dump site had not been overtly exposed to these chlorinated pesticides (37). In each of these studies, internal dose measurements substantially clarified exposure assessment and thus helped guide the risk management policy.

1,1,1-Trichloro-2,2-bis(*p*-chlorophenyl)ethane (DDT) in Triana, Alabama. In December 1978, publicity concerning high DDT residues in fish caught in Indian Creek, downstream from a defunct DDT manufacturing plant near Triana, Alabama, raised the question of increased exposure for the large number of people who ate locally caught fish. We analyzed the edible portions of several fish and the serum of 12 local residents, who ate several meals of fish weekly, for DDT and its metabolites, 1,1-dichloro-2, 2-bis(*p*-chlorophenyl)ethene (DDE) and 1,1-dichloro-2,2-bis(*p*-chlorophenyl) ethane (DDD). The average concentration of total DDT (i.e., DDT + DDE + DDD) in the fish was 226 ppm—the Food and Drug Administration's action level is 5 ppm; the 12 residents also had extremely high levels of these compounds, including the highest level ever reported in an individual. Subsequently, a community-wide study was conducted (*38*).

The 518 persons who registered for the study seemed ideal to model for exposure to DDT, because the primary environmental medium (food) and the approximate concentration of the pollutant in that medium were known, the primary route of entry (ingestion) was known, and the number of meals eaten per week of locally caught fish could be learned from questionnaires. Therefore, this situation offered a good opportunity to develop an exposure index in an environmental setting.

We measured DDT and its metabolites in 499 persons of a wide range of ages. The serum levels ranged from 0.6 to 2821 ppb with a geometric mean of 76.2 ppb, which is greater than five times the geometric mean found in the U.S. population in The Second National Health and Nutrition Examination Survey (NHANES II) (*37*). For the Triana study, the exposure index of concentration of the pollutant in the environmental medium (fish) and the amount of fish eaten per week were significantly related to internal dose; however, age was an even better predictor. These findings emphasize that for exposure to DDT and other lipophilic toxicants, age must be considered in the interpretation of measures of internal dose. In fact, PCBs were subsequently measured (*39*) in this population, and age was found to be the best predictor ($r = 0.56$) of the log PCB level (*40*). The Triana study demonstrated how measures of internal dose can help identify (1) predictors of exposure and (2) confounding factors that require adjustment in exposure assessment.

Occupational Dioxin Study. The National Institute for Occupational Safety and Health (NIOSH) of the CDC conducted a retrospective cohort study of 5172 workers at 12 plants in the United States that produced chemicals contaminated with 2,3,7,8-TCDD (*24*). The exposure index was based on the duration of exposure that was calculated from a careful and detailed review of each worker's occupational records. We then measured (*19*) the serum TCDD levels of 253 of these workers and a comparison group. The serum dioxin levels were significantly related to the exposure index (Pearson correlation coefficient of 0.72, $p < .0001$). On the basis of this correlation, the NIOSH exposure index can be used to assess any adverse health effects observed in its population. However, even in this study, the exposure index cannot be used to reliably predict the serum level of a given individual.

Conclusions

We have presented six exposure situations. In four of these (Operation Ranch Hand, Seveso, Triana, and the NIOSH Occupational Dioxin Study), we found highly exposed populations; in two studies (the U.S. Army ground troops in Vietnam and residents living near hazardous waste sites), we did not. The NIOSH Study represents an occupational setting in which the population is directly exposed to the source or the origin of the pollutant. The Operation Ranch Hand study is somewhat similar in that many of the men's military occupations caused them to handle the herbicide daily. In the NIOSH study, the exposure index predicted the internal dose for the population; in the Operation Ranch Hand study, it did not. Three other situations are more representative of chronic environmental exposure. In Triana, the primary route of the DDT's entry was ingestion of highly contaminated fish, whereas in the other two studies (U.S. Army ground troops in Vietnam and residents living near toxic sites), the pollutants' routes of entry could have been ingestion, inhalation, or skin contact, or a combination of these routes, and the environmental media could have been air, soil, food, or perhaps water. Finally, Seveso is an example of an acute environmental exposure.

The choice of methods to assess exposure depends on (1) intended use of the exposure data, (2) economic and logistic constraints, and (3) the availability and interpretability of measures of exposure (external dose, internal dose, or biologically effective dose) for the toxicant of interest. Environmental measurements of toxicant concentration in air, water, food, and soil will always be important for regulatory purposes. These external dose measurements should be correlated to measurements of the internal dose or biologically effective dose, whenever possible. The assessment of human exposure based on environmental monitoring alone must be done very cautiously.

For epidemiological studies of the relationship between exposure and health effects, every effort should be made to use the exposure measures closest to the health effect in the pathway shown in Figure 1. Usually, this is a measurement of internal dose (e.g., the concentration of the toxicant or its primary metabolite in blood or urine). If exposure indices based on environmental measurements or on questionnaire and history information are to be used to determine human exposure, rather than the direct measurements of internal dose or biologically effective dose, it is necessary to validate and calibrate the exposure index against direct measurements of internal dose in a subset of the exposed population, if at all possible. Calibration is needed so that investigators can compare the levels of exposure to a given toxicant in populations from different studies, such that a "high" level of exposure in one study is equivalent to a "high" level in another. Future efforts to characterize the pharmacokinetics of measures of internal dose and biologically effective dose are important. The pharmacokinetic studies are especially important for those toxicants that have short half-lives in humans. Because many environmental toxicants are ubiquitous, it is also necessary to know the normal concentration range of these toxicants in humans, stratified by age, gender, race, and so forth. We are currently determining these reference levels for 32 volatile organic compounds and 12 phenols and phenoxy acids in a subset of the Third National Health and Nutrition Examination Survey. We are also attempting to develop sensitive, specific, reliable, and affordable measures of internal dose and biologically effective dose, which, combined with sound epidemiology, will substantially improve human exposure assessment and human risk assessment.

References

1. Perera, F. P.; Weinstein, I. B. *J. Chron. Dis.* **1982**, *35*, 581–600.
2. Lioy, P. J. *Environ. Sci. Technol.* **1990**, *24*, 938–945.
3. Needham, L. L.; Pirkle, J. L.; Burse, V. W.; Patterson, D. G., Jr.; Holler, J. S. *J. Exp. Anal. Environ. Epidemiol.* **1992**, (Suppl. 1) 209–221.
4. Hulka, B. S.; Wilcosky, T. *Arch. Environ. Health* **1988**, *43*, 83–89.
5. Phillips, D. L.; Pirkle, J. L.; Burse, V. W.; Bernert, J. T, Jr.; Henderson, O.; Needham, L. L. *Arch. Environ. Contam. Toxicol.* **1989**, *18A*, 495–500.
6. American Conference of Governmental Industrial Hygienists *Threshold Limit Values for Chemical Substances and Physical Agents and Biological Exposure Indices (1991–1992);* American Conference of Governmental Industrial Hygienists: Cincinnati, OH.
7. Cline, R. E.; Hill, R. H., Jr.; Phillips, D. L.; Needham, L. L. *Arch. Environ. Contam. Toxicol.* **1989**, *18*, 475–481.
8. Needham, L. L.; Burse, V. W.; Price, H. A. *J. Assoc. Off. Anal. Chem.* **1981**, *64A*, 1131–1137.
9. Burse, V. W.; Needham, L. L.; Price, H. A.; Liddle, J. A.; Bayse, D. D. *J. Anal. Toxicol.* **1980**, *4*, 22–26.
10. Needham, L. L.; Cline, R. E.; Head, S. L.; Liddle, J. A. *J. Anal. Toxicol.* **1981**, *5B*, 283–286.
11. Needham, L. L.; Hill, R. H., Jr.; Orti, D. L.; Felver, M. E.; Liddle, J. A. *J. Chromatogr.* **1982**, *233*, 9–17.
12. Rutstein, D. D.; Veech, R. L.; Nickerson, R. J.; Felver, M. E.; Vernon, A. A.; Needham, L. L.; Kishore, P.; Thacker, S. B. *Lancet* **1983**, *2*, 534–537.
13. Needham, L. L.; Head, S. L.; Cline, R. E. *Anal. Lett.* **1984**, *17B(14)*, 282.
14. Patterson, D. G., Jr.; Holler, J. S.; Lapeza, C. R., Jr.; Alexander, L. R.; Groce, D. F.; O'Connor, R. C.; Smith, S. J.; Liddle, J. A.; Needham, L. L. *Anal. Chem.* **1986**, *58A*, 705–713.
15. Patterson, D. G., Jr.; Hoffman, R. E.; Needham, L. L.; Roberts, D. W.; Bagby, J. R.; Pirkle, J. L.; Falk, H.; Sampson, E. J.; Houk, V. N. *J. Am. Med. Assoc.* **1984**, *256*, 2683–2686.

16. Andrews, J. S., Jr.; Garrett, W. A., Jr.; Patterson, D. G., Jr.; Needham, L. L.; Roberts, D. W.; Bagby, J. R.; Anderson, J. E.; Hoffman, R. E.; Schramm, W. *Chemosphere* **1989**, *18*, 499–506.
17. Needham, L. L.; Patterson, D. G., Jr.; Alley, C. C.; Isaacs, S.; Green, V. E.; Andrews, J.; Sampson, E. J. *Chemosphere* **1987**, *16*, 2027–2031.
18. Needham, L. L.; Patterson, D. G., Jr.; Houk, V. N. In *Biological Basis for Risk Assessment of Dioxins and Related Compounds*; Gallo, M. A.; Scheuplein, R. J.; Van der Heiden, K. A., Eds.; Banbury Report; Cold Spring Harbor Laboratory: New York, 1991; Vol. 35, pp 229–257.
19. Patterson, D. G., Jr.; Hampton, L.; Lapeza, C. R., Jr.; Belser, W. T.; Green, V.; Alexander, L.; Needham, L. L. *Anal. Chem.* **1987**, *59*, 2000–2005.
20. Mocarelli, P.; Needham, L. L.; Marocchi, A.; Patterson, D. G., Jr.; Brambilla, P.; Gerthoux, P. M.; Meazza, L.; Carreri, V. *J. Toxicol. Environ. Health* **1991**, *32*, 357–366.
21. Holler, J. S.; Fast, D. M.; Hill, R. H., Jr.; Cardinali, F. L.; Todd, G. D.; McCraw, J. M.; Bailey, S. L.; Needham, L. L. *J. Anal. Toxicol.* **1989**, *13*, 152–157.
22. Hill, R. H., Jr.; To, T.; Holler, J. S.; Fast, D. M.; Smith, S. J.; Needham, L. L.; Binder, S. *Arch. Environ. Contam. Toxicol.* **1989**, *18*, 469–474.
23. Ashley, D. L.; Bonin, M. A.; Cardinali, F. L.; McCraw, J. M.; Holler, J. S.; Needham, L. L.; Patterson, D. G., Jr. *Anal. Chem.* **1992**, *64*, 1021–1029.
24. Fingerhut, M. A.; Halperin, W. E.; Marlow, D. A.; Piacitelli, L. A.; Honchar, P. A.; Sweeney, M. H.; Greife, A. L.; Dill, P. A.; Steenland, K.; Suruda, A. J. *N. Engl. J. Med.* **1991**, *324*, 212–218.
25. Hulka, B. S.; Margolin, B. H. *Am. J. Epidemiol.* **1992**, *135*, 200–209.
26. Pirkle, J. L.; Wolfe, W. H.; Patterson, D. G., Jr.; Needham, L. L.; Michalek, J. E.; Miner, J. C.; Peterson, M. R. *J. Toxicol. Environ. Health* **1989**, *27*, 165–171.
27. Michalek, J. E. *Appl. Ind. Hyg.* **1989**, *260*, 68–72.
28. Centers for Disease Control and Prevention Veterans Health Studies. *J. Am. Med. Assoc.* **1988**, *260*, 1249–1254.
29. Burse, V. W.; Needham, L. L.; Korver, M. P.; Lapeza, C. R., Jr.; Liddle, J. A.; Bayse, D. D. *J. Assoc. Off. Anal. Chem.* **1983**, *66*, 32–39.
30. Mortensen, B. K.; Anderson, J. E.; Burse, V. W.; Garrett, W. A.; Kotlovker, D.; Walter, G. *Exposure Study of Persons Possibly Exposed to Polychlorinated Biphenyls in Paoli, Pennsylvania*. Final Report—Technical Assistance to the Chester County, PA, Health Department; Agency for Toxic Substances and Disease Registry: Atlanta, GA, November 1987.
31. Phillips, D. L.; Smith, A. B.; Burse, V. W.; Steele, G. K.; Needham, L. L.; Hannon, W. H. *Arch. Environ. Health* **1989**, *44B*, 351–354.
32. Burse, V. W.; Groce, D. F.; Korver, M. P.; Caudill, S. P.; McClure, P. C.; Lapeza, C. R., Jr.; Head, S. L.; Schilling, R. J.; Farrar, J. A.; Ostrowski, S. R.; Mortensen, B. K.; Maher, J. P.; Russell, H. L.; Sievila, S. *J. Assoc. Off. Anal. Chem.* **1991**, *74*, 579–586.
33. Stehr-Green, P. A.; Burse, V. W.; Welty, E. *Arch. Environ. Health* **1988**, *43*, 420–424.
34. Miller, D. T.; Condon, S. K.; Kutzner, S.; Phillips, D. L.; Krueger, E.; Timperi, R.; Burse, V. W.; Cutler, J.; Gute, D. *Arch. Environ. Contam. Toxicol.* **1991**, *20*, 410–416.
35. Telles, N. *The New Bedford PCB Study—Preliminary Findings: March 23, 1982*; Massachusetts Department of Public Health: Boston, MA, 1982.
36. Burse, V. W.; Head, S. L.; Korver, M. P.; McClure, P. C.; Donahue, J. F.; Needham, L. L. *J. Anal. Toxicol.* **1990**, *14*, 137–142.

37. Andrews, J. S., Jr.; Anderson, J. E.; Rifenburg, J. A.; Rowley, D. L.; Kahn, S. E.; Hudson, R. F.; Burse, V. W.; Needham, L. L.; LaChapelle, N. C. *Health Effects Study of Residents near the Hollywood Dumpsite, Memphis, Tennessee*; Final Report Cooperative Agreement No. U61–CCU400608–01; Memphis and Shelby County Health Department and the Center for Environmental Health and Injury Control, Centers for Disease Control and Prevention: Atlanta, GA, 1988.
38. Kreiss, K.; Zack, M. M.; Kimbrough, R.D.; Needham, L. L.; Smrek, A. L.; Jones, B. T. *J. Am. Med. Assoc.* **1981**, *245*, 1926–1930.
39. Needham, L. L.; Smrek, A. L.; Head, S. L.; Burse, V. W.; Liddle, J. A. *Anal. Chem.* **1980**, *52*, 2227–2229.
40. Kreiss, K.; Zack, M. M.; Kimbrough, R. D.; Needham, L. L.; Smrek, A. L.; Jones, B. T. *J. Am. Med. Assoc.* **1981**, *245*, 2505–2509.

RECEIVED for review September 3, 1992. ACCEPTED revised manuscript February 4, 1993.

11

Research Strategy for Assessing Human Health Risks from Exposure to DNA-Reactive Chemicals
1,3-Butadiene as a Case Study

James A. Bond, Leslie Recio, and Roger O. McClellan

Chemical Industry Institute of Toxicology, P.O. Box 12137, Research Triangle Park, NC 27709

> *The interaction of chemicals or their metabolites with DNA is a major factor in chemical carcinogenesis. One potential strategy for more accurately estimating human health risks from exposure to DNA-reactive chemicals is to develop research that will improve our understanding of the mechanisms of action of chemicals within an exposure → tissue dose → cancer response paradigm. The development of quantitative linkages between exposure and response, which are based on biologically plausible mechanisms of action at exposure levels likely to be encountered by people, will significantly improve the risk assessments for human exposures to DNA-reactive chemicals. 1,3-Butadiene (BD) represents an interesting case study in which the foregoing research strategy is providing data that are critical for understanding interspecies differences in responses to BD. BD is a carcinogen in rats and mice; mice are more sensitive than rats. It is not known whether BD poses a carcinogenic risk for humans. BD requires metabolic activation to DNA-reactive epoxides that can bind to DNA to initiate a series of events leading to tumor formation. Species differences in activation and detoxication need to be considered in developing human risk estimates for BD. BD activation-to-detoxication ratios are markedly different for mouse (74), rat (6), and human (6) liver tissues. The differences in the ratios between mice and rats are consistent with the higher carcinogenic sensitivity of mice to BD, compared with rats. Additional data, which were developed by using a transgenic mouse model system, have indicated that BD induces in vivo mutations in target tissues. These data coupled with the BD metabolism data in mice, rats, and humans form the basis of a mechanistically based BD tissue dosimetry model in rats and mice*

0065–2393/94/0241–0137$08.00/0
© 1994 American Chemical Society

that can now serve as the basis for a human dosimetry model. This dosimetry model enables key extrapolations from high- to low-dose exposures and between species and will be employed in a human risk assessment for BD.

THE INTERACTION OF CHEMICALS OR THEIR METABOLITES WITH DNA is a major factor in carcinogenesis of a number of chemicals (1, 2, and references therein). A central hypothesis for the mechanism of action of many DNA-reactive chemical carcinogens is that the generation of DNA damage coupled with subsequent cell replication can result in mutations in specific genes that play a key role in the carcinogenic process. However, not all types of damage to DNA, whether in the form of DNA adducts, cross-links, or base deletions, will have the same impact on the replicating cell. For example, some abundant and stable DNA adducts are excellent markers of exposure but may be poor markers of risk, because they are not promutagenic. DNA adducts may occur at sites that are not essential for cell function and thus would have little impact on cancer development. Additionally, the ability of a cell to repair the damaged DNA and the rate and fidelity of the repair will have important consequences for whether the damage is fixed and ultimately results in a mutational event. Thus, knowledge of how chemicals interact with DNA, the probability of that interaction inducing a specific mutation, and the subsequent probability of tumor formation is critical for the development of biologically based models to predict the human health consequences for DNA-reactive chemicals.

A major goal of the Chemical Industry Institute of Toxicology (CIIT) DNA-reactive chemical research strategy is the development of a biologically based mechanistic approach to assessing human cancer risks for exposure to these types of chemicals. The research involves the development of biologically based descriptions of dosimetry (i.e., metabolism and hemoglobin- and DNA-adduct formation) and incorporation of mechanisms of chemically induced mutation into a biologically based framework for tumor formation. Several components to the DNA-reactive chemical research strategy are used at CIIT. These components, illustrated in Figure 1, include biomarkers, dosimetry, DNA interaction, mutation induction, cell proliferation, and tumor induction. Data developed from biomarker, dosimetry, and DNA interaction studies serve as critical input for the dosimetry models that are ultimately developed. Data developed from DNA interaction studies as well as mutation induction, cell proliferation, and tumor induction studies, coupled with the dosimetry models, can support the development of biological response models.

A recommended research approach for studying DNA-reactive chemicals includes experiments that quantitatively describe the translocation of material from the external environment to critical target sites on DNA and

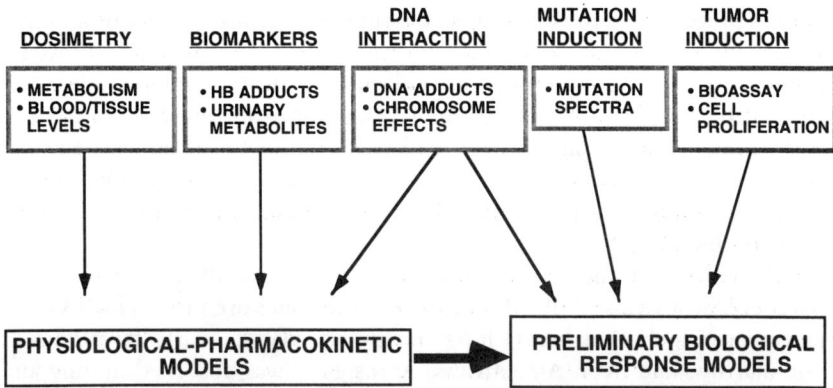

Figure 1. Important components of a DNA-reactive chemical research strategy designed with a biologically based mechanistic approach for assessing human cancer risks for exposure to these types of chemicals.

the role of the damaged DNA in increasing the likelihood of mutation. These experiments involve such dynamic processes as absorption, distribution, and in the case of nonreactive chemicals, metabolism to DNA-reactive and stable nontoxic compounds. The formulation of quantitative descriptions of the metabolic pathways involved in the production of DNA-reactive metabolites should be a key component of the approach. These descriptions include both activation and detoxication pathways, because the ultimate concentration of the DNA-reactive species at the target site depends on the balance between these pathways. It is likely that both exposure concentration and exposure rate changes can affect the balance between activation and detoxication processes. The rates of DNA damage (i.e., DNA adduct formation and DNA strand breaks) and repair as well as the time course for formation and persistence of adducts are also major considerations. Additionally, knowledge of cell replication can provide a critical linkage between interaction of chemicals with DNA and the ability of the damaged DNA to induce mutational events and subsequent tumor development. An evaluation of the use of macromolecular adducts formed from reactive chemicals as monitors of exposure, in which the use of the biomarkers as an indicator of tissue dose is particularly emphasized, is another important consideration. The macromolecular adducts should include DNA adducts and hemoglobin adducts. Ultimately, this research strategy will specify the relationships among different biomarkers of exposure with respect to their relative levels and time of appearance after exposure to DNA-reactive chemicals.

BD as a Case Study

BD represents an interesting case study in which the foregoing research strategy is providing data that are critical for understanding interspecies differences in responses to BD. BD, a monomer widely used in the production of synthetic rubber and other resins, exhibits low acute inhalation toxicity in rodents ($LC_{50} < 120,000$ ppm) (3) and causes sensory irritation in humans at concentrations of 2000–8000 ppm (4). Occupational exposures to BD can result during production, storage, and transport of the chemical. BD has been detected in cigarette smoke (5) and automobile exhaust (6) and is currently listed as one of the 189 hazardous air pollutants in the 1990 Clean Air Act Amendments (7).

BD induces tumors at multiple organ sites in B6C3F1 mice and Sprague–Dawley rats (8–10). In mice, several tissues are targets for BD carcinogenicity, including heart, lung, mammary gland, ovary, forestomach, bone marrow, and liver. An activated K-ras gene was observed in lung and liver tumors and lymphomas of B6C3F1 mice exposed to BD (11). The striking aspect of BD-induced carcinogenicity is the high sensitivity of mice to BD. Tumors were observed in mice at concentrations as low as 6 ppm, and steep concentration–response curves were evident for several tumors. The tumor sites in rats are different and include the thyroid, mammary gland, Zymbal gland (auditory subaceous gland), uterus, testis, and pancreas. Rats exhibit a relatively low sensitivity to BD, because tumors occur at BD concentrations (1000–8000 ppm) nearly three orders of magnitude higher than in mice.

BD displays mutagenic activity in *Salmonella typhimurium*, but only in the presence of hepatic 9000-g supernatant (12–14), which indicates that BD is not mutagenic but that its metabolites, possibly 1,2-epoxybut-3-ene [butadiene monoepoxide (BMO)] and 1,2:3,4-diepoxybutane [butadiene diepoxide (BDE)], are responsible for the observed bacterial mutagenicity. BD is also genotoxic in vivo, inducing chromosome aberrations and sister chromatid exchanges in bone marrow cells and micronucleus formation in peripheral blood of male B6C3F1 mice (15,16) but not in Sprague–Dawley rats (16).

Studies on the induction of BD-induced mutational events have primarily been carried out in cell culture systems, and extrapolation of these data to in vivo genetic effects is difficult. Therefore, in vivo approaches to determine mutation induction in somatic cells of tissues need to be considered. Transgenic mice constructed with mutational target genes can permit the study of dose–response and the molecular basis of mutation in tissues from animals with defined carcinogenic exposures.

Certain transgenic mouse models are in vivo shuttle vector systems (17) that use bacterial genes (e.g., *lacI* or *lacZ*) as mutational targets inserted within the λ-phage genome that is integrated into the mouse genome. The induction of mutations can be determined with respect to exposure, dose-to-target tissues, and DNA adduct levels, thereby providing a link between in vivo exposures and mutagenic events. The mutated transgene can be sequenced, thereby providing the opportunity for assessment of mutational specificity. This provides a unique opportunity to determine the induction of BD mutations in mice that will reflect the integrated in vivo pharmacokinetic and biotransformation processes that ultimately result in BD-derived mutagenic metabolites.

Three enzymes appear to play major roles in the overall metabolism of BD, cytochrome P-450–dependent monooxygenases, epoxide hydrolases, and glutathione (GSH) S-transferases. In addition to enzymic reactions, BD epoxides may nonenzymatically hydrolyze or conjugate with GSH. Presently, insufficient data quantitatively describe the contribution of these various pathways in Sprague–Dawley rats and B6C3F1 mice, species susceptible to BD-induced carcinogenesis, or humans, a species potentially at risk for BD. Because one or more of the BD epoxides may play a role in the carcinogenicity of BD, a quantitative determination of the balance of activation (i.e., epoxide formation) and inactivation (i.e., epoxide hydrolysis or conjugation) is essential for improving our understanding and assessment of human risk following exposure to BD.

This chapter describes quantitative species differences in the oxidation of BD and BMO by cytochrome P450–dependent monooxygenases and in the inactivation of BMO by epoxide hydrolases and glutathione S-transferases found by using microsomal and cytosolic preparations of livers obtained from Sprague–Dawley rats, B6C3F1 mice, transgenic mice (CD2F1), and humans. In addition, transgenic mice were used to assess the capacity of BD to induce in vivo mutations. The data from these studies provide a quantitative description of the relative contributions of the various pathways of BD metabolism in three animal species, including humans, and the role that metabolism plays in forming mutagenic DNA-reactive BD metabolites. Data developed on BD metabolism can serve as the basis for the development of a BD dosimetry model. The BD dosimetry model will enable key extrapolations from high- to low-dose exposures and between species and will be predictive of target tissue levels of reactive BD metabolites. The dosimetry model can by employed in a human risk assessment for BD.

Materials and Methods

Animals. Male Sprague–Dawley (CD) rats (9–10 weeks) and male B6C3F1 mice (9–10 weeks) were obtained from Charles River Laboratories (Raleigh, NC). The MutaMouse (MM) transgenic strain [BALB/c × DBA/2 (CD2F1)] of mice (6 weeks) was purchased from Hazleton Research Products, Inc. The construction of the shuttle vector in λ-phage (λgt 10 *lacZ*), with the inserted bacterial target gene for mutagenesis (*lacZ*) and production of the transgenic mice, is described in detail elsewhere (*18*). All animals were determined to be free from viral infection and were acclimated for at least 2 weeks prior to use. Animals were fed with standard diet (NIH-07; Zeigler Brothers, Gardners, PA) and received water ad libitum. They were maintained on a 12-h light–dark cycle, beginning at 0700, and housed at 22 ± 2 °C and 55 ± 5% relative humidity.

Preparation of Liver Microsomes for Metabolism Studies. Rodents were euthanized by either sodium pentobarbital or CO_2 asphyxiation, and the livers were excised, frozen in liquid nitrogen, and stored at −80 °C. Livers were slowly thawed while on ice, weighed, cut into pieces, and homogenized with 4 volumes of isotonic KCl–Tris buffer with six passes of a Teflon-glass homogenizer (1100 revolutions per minute Braun). Microsomes and cytosol were prepared as described previously (19).

Samples of liver ($n = 12$) from trauma victims (Tennessee Donor Services, Nashville, TN) were used in these experiments. Samples were frozen (−80 °C) and remained frozen during shipment. Tissues were slowly thawed on ice and homogenized, and microsomes and cytosol were isolated as described previously (19). Typically, all 12 samples were used for the initial assessment of reaction rates for an enzyme-catalyzed reaction. The 12 samples were then rank-ordered by enzyme activity, and the three samples representing the highest, lowest, and median enzyme activities were used for the enzyme kinetic experiments. For determination of BD oxidation kinetics, all 12 liver samples were used.

Protein content was determined by using a modified micro-Lowry method (20). Cytochrome P450 content was estimated spectrally (21).

Cytochrome P450-Dependent Metabolism of BD. Microsomes were diluted with 0.1-M phosphate buffer (pH 7.4) to the desired protein concentration. One milliliter of diluted microsomes was placed in each 10-mL vial (Hypo Vial, Pierce Chemical Company, Rockford, IL), which was sealed with Tuf-Bond septa (Pierce Chemical Company, Rockford, IL). Vials were shaken at 37 °C in a Dubnoff metabolic shaking incubator. After a 10-min preequilibration period, BD (600–25,000 ppm) in a gas-tight syringe (Hamilton Company, Reno, NV) was injected into the sealed, airtight vials. After an additional 10-min preincubation, the enzymic reaction was started by adding 100 μL reduced nicotinamide adenine dinucleotide phosphate (NADPH) (4 μmol). Preliminary experiments indicated that this preincubation period was essential for equilibration of BD between the liquid and gas phases.

Air samples (100 μL) were withdrawn from the headspace (i.e., gas phase) of the vial at 5-min intervals (up to 45 min) and analyzed by gas chromatography with a Hewlett-Packard gas chromatograph (HP 4890A Series II) equipped with a flame ionization detector and a 7-ft × 1/8-in stainless-steel column filled with diphenylphenylene oxide (Tenax) 35–60 mesh. The carrier gas was helium (18.5 mL/min) and the detector, column, and injector temperatures were 250, 130, and 200 °C, respectively. Under these conditions, the retention times of BD and BMO were 0.80 min and 2.20

min, respectively. Calibration curves of BD and BMO were linear in the concentration ranges of 10–55,000 ppm and 1–340 ppm, respectively.

Enzyme-Mediated Hydrolysis of BMO. Enzyme-mediated hydrolysis of BMO by epoxide hydrolase was measured exactly as described for the cytochrome P450-dependent reactions, except that BMO was used as the substrate for the reaction and NADPH was not added. NADPH was omitted from the reactions to prevent enzymic oxidation of BMO. To assess the nonenzymic hydrolysis of BMO, reactions were carried out in the presence of heat-inactivated tissue.

Enzyme-Mediated Conjugation of BMO with GSH. The reaction mixtures (0.1 mL) containing different BMO concentrations (2.5, 7.4, 17.4, 37.1, and 74.5 mM), cytosolic protein (1.0 mg/mL), and GSH (10 mM) were placed in Screw Cap Septum vials (Pierce Chemical Co., Rockford, IL) and sealed with Tuf-Bond septa. ^{3}H-GSH (40 μCi/mL) was added to the vials, and after a 10-min incubation at 37 °C, 0.1 mL cooled methanol was added, and the precipitated proteins were centrifuged for 1 min at 12,500 g. Reactions were linear with time for at least 10 min. An aliquot of supernatant (10 μL) was chromatographed by high-pressure liquid chromatography (HPLC) (Waters 510 pumps, Milford, MA). Metabolites were separated on a 5-μm C_{18} Ultrashere column (250 × 4.6 (i.d.) mm; Beckman) and the absorbance of the effluent was monitored by UV detection (Waters 486 Tunable Absorbance Detector, λ = 263 nm). Metabolites were quantitated by monitoring the radioactivity of the eluant with a Packard FlowOne Beta in-line radioflow detector (Packard, Downers Grove, IL). Analysis was performed under isocratic conditions of methanol:0.1% trifluoroacetic acid (15:85) with a 1.0 mL/min flow rate. The retention times for GSH, oxidized glutathione (GSSG), and the two major GSH conjugates, S-(1-hydroxy-3-buten-2-yl)glutathione and S-(2-hydroxy-3-buten-1-yl)glutathione, were 3.8, 4.5, 7.8, and 8.5 min, respectively. Data on rates of formation of GSH conjugates that are reported represent the sum of the individual rates of formation for each of the conjugates. Recovery of GSH, GSSG, and metabolites from the column was greater than 99%.

Experimental Design for Mutagenicity Studies. The mutagenicity experiment utilized three groups of animals with five MM per group. The BD group was exposed to 625 ppm BD for 5 consecutive days for 6 h/day. This exposure concentration was selected because previous studies indicated that exposure of B6C3F1 mice to 625 ppm BD for up to 60 weeks resulted in significant elevation of tumor incidences in a number of organs (8). The two remaining groups of animals consisted of air controls that were housed in chambers similar to those of the BD-exposed group and a group of animals administered N-ethyl-N-nitrosourea (ENU) at 250 mg/kg (intraperitoneal injection dissolved in dimethyl sulfoxide) as a positive control. Animals were housed in individual wire mesh cages inside a 1-m^3 Hinners-style chamber (22). During the 6 h of exposure to air or BD, the animals were without food but had free access to water. Air flow in the chamber was maintained within 10% of 15 air changes per hour.

BD exposure levels in the chamber were monitored by infrared spectroscopy using a Foxboro Wilkes Miran IA IR (Norwalk, CT). BD was generated directly from a liquid–gas storage tank operated at approximately 24 pounds per square inch. The vapor was passed through a calibrated flow control device prior to mixing with the high-efficiency particulate air-filtered dilution air supplying the exposure chamber. The BD exposure concentrations throughout the chamber were within 5% of the target. Relative humidity in the chamber was maintained at 50 ± 10%, and temperature was maintained at 22 ± 2 °C. The five control mice were similarly housed in a second chamber in the same room. They were exposed to clean air of the same temperature, relative humidity, and air flow as that delivered to the BD-exposed animals.

Animal Necropsy and Tissue Collection. MM were killed by an overdose of sodium pentobarbital and exsanguinated by cardiac puncture, and the blood was collected in heparinized syringes. The liver and lungs excised from the animals were immediately frozen in liquid nitrogen and subsequently stored frozen at −80 °C. Bone marrow was removed from the tibias and femurs by rinsing with Dulbecco's phosphate buffered saline (GIBCO, Grand Island, NY). The cell suspension was placed in a microcentrifuge tube, and the bone marrow cells were pelleted in a microcentrifuge at maximum speed for 30–45 sec. The supernatant was discarded; the cell pellet was frozen in liquid nitrogen and then stored at −80 °C until the DNA was extracted.

Bacterial Strains and Media. The following bacterial strains ($mcrA^-$ and $mcrbB^-$) were used: *Escherichia coli* C/*lacZ*$^-$ *lacI*- (tetracycline, ampicillin) and *E. coli* C-*PLK*/*lacZ*$^-$ (Stratagene, La Jolla, CA). Standard bacterial Luria–Bertani (LB) medium (21) was used for bacterial growth and plating. For visualization of plaques, 5-bromo-4-chloro-3-indolyl-β-D-galactoside (X-gal) (0.45 mg/mL) was included in the medium.

DNA Extraction. Frozen tissues were thawed in approximately 5 mL of lysis buffer (10-mM tris-HCl pH 8.0, 150 mM NaCl, 20-mM ethylenediaminetetraacetic acid pH 8.0) and carefully homogenized using a handheld Dounce homogenizer. DNA was extracted from tissues by using standard techniques (24). The concentration of DNA was determined by UV absorption.

Packaging of λ-Phage DNA from the Mouse Genome and Phage Titering. The λ-phage genome was "packaged" from the mouse genome by using Giga-Pack Gold packaging extract (Stratagene, La Jolla, CA). Approximately 7.5 μg of mouse genomic DNA (1.5 mg/mL) was used in each packaging reaction. Packaging reactions were initiated according to the manufacturer's directions, using a 3-h incubation at 37 °C. The reaction was terminated by dilution with 1.0 mL of SM buffer (50-mM Tris-HCl, pH 7.5, 10-mM $MgSO_4$, 100-mM NaCl, 0.01% gelatin).

The packaging reaction (2.5 μL) was adsorbed onto a log growth culture of *E. coli C/lacZ*$^-$ by gently mixing and incubating at room temperature for 20 min. The mixture was added to LB medium with agarose (plus X-gal) and poured onto a 150-mm plate. The plates were incubated at 37 °C overnight, and the number of plaques per plate were counted. The amount of packaged DNA required to produce approximately 1500 plaques was calculated, as was the number of plaque-forming units (pfu) per microgram of DNA used in the packaging reaction. All packaging reactions yielded more than 10,000 pfu/μg DNA.

Determination of *lacZ*$^-$ Mutant Frequency. To determine the *lacZ*$^-$ mutant frequency in bone marrow, lung, and liver tissue samples, DNA was independently extracted from each tissue for each animal. The DNA samples for each animal in the same exposure group (five per each exposure group) were packaged independently into λ-phage. After the determination of the pfu for each individual DNA sample, an equal number of pfu from each individual animal within an exposure group were combined to determine the *lacZ*$^-$ mutant frequency. Therefore, a single *lacZ*$^-$ mutant frequency was determined for each tissue for each exposure group. A total of 1×10^5 λ-phage plaques were examined from the bone marrow samples, and in lung and liver samples, approximately 5×10^5 λ-phage plaques were examined.

A log growth culture of *E. coli C/lacZ*$^-$ (5 mL) at 2.5×10^8 bacteria/mL and a titered volume of packaging reaction (DNA from five animals) that would produce 4500 plaques were combined in a 50-mL culture tube and incubated at room temperature for 20 min. LB medium with agarose (plus X-gal) was added to each tube, which was then pipetted onto three 150-mm plates. The plates were incubated at 37 °C overnight.

The total number of blue plaques (*lacZ*$^+$) was estimated by counting a defined, one-tenth area of three plates and extrapolating the mean to the full plate and the number of plates for the entire plating. The number of *lacZ*$^-$ mutant clear or light blue plaques were counted and "picked" for confirmation by replating on LB medium with agarose plus X-gal. The *lacZ*$^-$ mutant frequency was calculated by dividing the total number of confirmed mutant plaques by the total phage population analyzed and was expressed as $lacZ^-$ mutants/10^5 plaques. The number of observed mutant plaques within each tissue group was used to determine statistical significance relative to the tissues from the air control group, assuming a Poisson distribution.

Results

Mutant Frequency in Different Tissues of Mice following Exposure to BD. The $lacZ^-$ mutant frequency in three tissues of mice exposed to 619 ± 3.2 ppm BD (mean ± standard deviation [SD]) and ENU was determined 14 days following the last exposure. The number of $lacZ^-$ mutant plaques observed, and the $lacZ^-$ mutant frequency for each tissue is shown in Table I. The background $lacZ^-$ mutant frequency in the tissues examined from control mice ranged from 2 to 4 mutants/10^5 plaques. The positive control, ENU, was mutagenic in all tissues examined. The $lacZ^-$ mutant frequency in the BD-exposed mice did not show a significant increase in the bone marrow or liver. However, there was a significant increase in mutant frequency in the lungs of exposed mice ($p < .001$).

Microsomal Metabolism of BD. The initial rate of BD oxidation in liver microsomes was linear with protein content over the range of 1.0–9.0

Table I. $lacZ^-$ Mutant Frequency in Bone Marrow, Lung, and Liver Samples from BD Exposed Animals and Controls[a]

Tissue	No. of Mutants	Mutant Frequency ($\times 10^{-5}$)
Bone marrow		
Air control	3	3.0
BD	2	1.8
N-Ethyl-N-nitrosourea (ENU)-exposed	71	73.1[b]
Lung		
Air control	29	4.4
BD	59	9.1[b]
ENU-exposed	88	16.2[b]
Liver		
Air control	12	2.4
BD	18	3.1
ENU-exposed	71	13.0[b]

[a] BD-exposed mice were exposed to 625 ppm BD (6 h/day for 5 consecutive days). Controls were exposed to air only for the same duration as the BD-exposed mice. Mice dosed with ENU received a single intraperitoneal injection of 250 mg ENU/kg. All mice were killed 14 days after the last inhalation exposure or injection for mutant frequency analysis.
[b] Significantly greater ($p < .001$) than air control for each tissue by Poisson analysis of the number of mutant plaques relative to air controls.

mg protein/mL and with time for up to 30 min (data not shown). For the enzyme-mediated reactions using liver microsomes, both the disappearance of BD from the gas phase and the appearance of BMO in the gas phase were measured. The previously determined parameters of BMO metabolism (*see* further on) were incorporated into equations (*19*) used to describe the in vitro system. The equations were then used to estimate the Michaelis–Menten constants for BD oxidation as described by Csanady et al. (*19*). The model was found to adequately describe both BD disappearance from the headspace and BMO appearance in the headspace of reaction vials containing liver microsomes from all species. The Michaelis–Menten parameters are listed in Table II.

The maximum velocity (V_{max}) for BD oxidation to BMO for human liver microsomes was about one-half that observed with B6C3F1 mouse liver microsomes and twofold higher than in rat liver microsomes (Table II). The V_{max} for the reaction was about 1.5-fold higher in B6C3F1 mouse liver microsomes, compared with MM liver microsomes. The apparent Michaelis constants (K_ms) for BD oxidation in liver microsomes from humans, Sprague–Dawley rats, B6C3F1 mice, and MM were similar and ranged from 2 to 5 μM. There were some striking differences in the calculated V_{max}/K_m (*25*). The calculated V_{max}/K_m for BD oxidation in B6C3F1 mouse liver microsomes was sixfold greater than for rat and human liver microsomes, and about 50% higher than MM liver microsomes.

Microsomal Metabolism of BMO. Initial rates of enzymic hydrolysis of BMO in liver microsomes were linear with protein content over the range of 0.2–10 mg/mL and with time for up to 30 min (data not shown). Enzyme-mediated hydrolysis of BMO was not detected in cytosolic fractions of liver from any of the species. All 12 of the human liver samples were assessed for their ability to metabolize BMO using one initial BMO concentration (100 ppm). The normalized first-order rate constants for the 12 liver samples, which varied between 0.020 and 0.068 min^{-1} mg $protein^{-1}$, were then rank-ordered, and three samples (high, median, and low activity) were selected for detailed studies of BMO kinetics (*see* further on). In all cases,

Table II. Kinetic Constants for the Oxidation of BD to Butadiene Monoepoxide[a]

Liver Microsomes	V_{max} (nmol/mg protein/min)	K_M (mM)	V_{max}/K_M[b]
Humans	1.2 ± 0.6	0.005 ± 0.003	230
B6C3F1 mice	2.6 ± 0.06	0.002 ± 0.0002	1295
Sprague-Dawley rats	0.6 ± 0.1	0.004 ± 0.0002	157
MutaMouse	1.7 ± 0.1	0.002 ± 0.0002	850

[a] Values are mean ± standard deviation derived from model simulations (*19*).
[b] Values are in units of L·nmol·(min·mg protein·mmol)$^{-1}$.

results of model simulations (19) adequately described the decline in BMO concentration in the headspace due to enzymic hydrolysis (data not shown).

Table III shows the Michaelis–Menten constants (K_M, V_{max}) for BMO hydrolysis by human, rat, and mouse (B6C3F1 and MM) liver microsomes. In the three human liver samples used for the detailed kinetic experiments, V_{max} ranged from 9 to 60 nmol/mg protein/min, and apparent K_Ms ranged from 0.2 to 1.6 mM. In contrast, V_{max} determined from reactions with rodent liver microsomes was one-half or less of that measured in human liver samples. Interestingly, the V_{max} in B6C3F1 mice liver microsomes for BMO hydrolysis was about fivefold higher than in MM liver microsomes. Apparent K_Ms for B6C3F1 mice were approximately seven- to eightfold higher than the K_Ms observed in rats and five times higher than MM. A comparison of the calculated V_{max}/K_M reveals some striking differences across species (Table III). For example, for the three human liver samples, calculated V_{max}/K_Ms ranged from 32 to 38, whereas in rodents, calculated V_{max}/K_Ms ranged from 4 to 10.

Conjugation of BMO with GSH. Enzyme-mediated conjugation of GSH with BMO in human and rodent liver cytosol could best be described by Michaelis–Menten kinetics (Table IV). The V_{max} in B6C3F1 mouse liver cytosolic fractions was about twofold higher than in rat liver cytosol (Table IV). Conjugation of BMO with GSH in MM liver cytosol was not measured in these experiments. Only one of the two human liver samples analyzed displayed Michaelis–Menten kinetics and had a V_{max} of one-fifth to one-tenth of that observed in rodents. Mouse liver cytosolic apparent K_Ms were about threefold higher than that of rats and humans. The calculated V_{max}/K_M in rodent liver cytosol was about fourfold higher than the calculated V_{max}/K_M for the one human sample.

Discussion. An understanding of the mechanisms by which BD induces tumors in rats and mice is essential for extrapolating to humans, a

Table III. Kinetic Constants for the Hydrolysis of Butadiene Monoepoxide[a]

Liver Microsomes	V_{max} (nmol/mg protein/min)	K_M (mM)	V_{max}/K_M[c]
Humans[b] (high)	58.1 ± 4.0	1.65 ± 0.12	35
(median)	18.5 ± 4.1	0.58 ± 0.15	32
(low)	9.2 ± 2.2	0.24 ± 0.10	38
B6C3F1 mice	5.8 ± 0.3	1.59 ± 0.03	3.6
Sprague–Dawley rats	2.5 ± 0.05	0.26 ± 0.01	9.5
MutaMouse	1.1 ± 0.3	0.29 ± 0.10	3.8

[a]Values are mean ± standard deviation derived from model simulations (19).
[b]Three of the 12 human liver samples were used for the kinetic analyses. Selection of the samples were as described in the materials and methods section.
[c]Values are in units of $nmol \cdot L \cdot (min \cdot mg\ protein \cdot mmol)^{-1}$.

Table IV. Kinetic Constants for Conjugation of Butadiene Monoepoxide with Glutathione[a]

Liver Cytosol	V_{max} (nmol/mg protein/min)	K_M (mM)	V_{max}/K_M[c]
Humans[b]	45.1 ± 5.8	10.4 ± 1.04	4.3
B6C3F1 mice	500 ± 64	35.3 ± 6.2	14
Sprague–Dawley rats	241 ± 3	13.8 ± 0.3	17
MutaMouse	—[d]	—[d]	—[d]

[a]Values are mean ± standard deviation (SD) derived from model simulations.
[b]Two of the 12 human liver samples were used for kinetic analyses. Selection of the samples was as described in the materials and methods section. One of the samples displayed Michaelis–Menten kinetics and in the other sample, the reaction was best described by a rate constant of $(2.56 \pm 0.22) \times 10^{-4}$ L/mmol/min/mg protein (mean ± SD).
[c]Values are in units of $L \cdot nmol \cdot (min \cdot mg\ protein \cdot mmol)^{-1}$.
[d]Not measured.

species for which BD carcinogenic potency is at present unknown. It is likely that one of the critical biochemical determinants of BD-induced carcinogenicity is the extent to which BD is activated to epoxide metabolites that can react with DNA to ultimately induce mutations. The induction of mutations in tissues of a transgenic mouse is a novel approach to determine in vivo mutation induction following exposure to carcinogens. The induction of mutations can be studied within the context of the endogenous pharmacokinetic and biotransformation processes that determine the tissue levels of reactive metabolites that can interact with and alter the cellular DNA in various tissues.

BD is a mutagenic carcinogen that exhibits species differences in susceptibility; mice are more susceptible to the genotoxic and carcinogenic effects of BD than are rats. Studies on the in vitro and in vivo metabolism of BD indicate that mice produce more BMO and BDE than do rats (19, 26–28). These data indicate that the biotransformation of BD to two reactive metabolites, BMO and BDE, exhibits species differences and is likely a critical determinant of the genotoxicity and carcinogenicity of BD. The relative balance between the in vivo bioactivation and detoxification pathways of BD determines tissue concentrations of the ultimate mutagenic species, the quantities that interact with DNA, and ultimately the type and amount of promutagenic lesions in the DNA of exposed animals. Because in vitro systems cannot mimic certain of these in vivo processes, it is essential that methods for the in vivo evaluation of mutagenicity be considered.

Exposure of the MM strain of transgenic mouse to 625 ppm of BD for 5 consecutive days (6 h/day) followed by a 14-day expression period did not result in significant mutagenicity in liver and bone marrow. However, significant mutagenicity was observed in the lung, a target organ for the carcinogenic effect of BD in B6C3F1 mice at 6.25 ppm. Specific mutation of the K-ras oncogene has also been detected in lung adenocarcinomas in BD-exposed mice (*11*). Because no carcinogenicity data in the MM strain of mouse exist as they do for the B6C3F1 mouse, it is difficult, at present, to relate the relatively small increases in mutagenicity observed in the present studies to another biological end point (e.g., tumor formation). Studies are in progress to evaluate the mutagenicity of BD using a B6C3F1 mouse mutation system (*29*). Exposure of B6C3F1 mice to BD resulted in significant genotoxicity as determined by micronucleus (MN), sister chromatid exchanges (SCE), and chromosomal aberration (CA) analysis (*15,16*). Significant levels of MN (sixfold above control levels), and SCE were observed in B6C3F1 mice following 2-day exposures (6 h/day; nose only) at levels of 100 ppm BD and higher (*16*). In the same study, exposure of rats to 100–10,000 ppm (6 h/day for 2 days) did not result in significant increases in SCE or MN. Exposure of B6C3F1 mice to 6.25, 62.5, and 625 ppm BD 6 h/day for 10 days resulted in a significant dose-dependent increase in SCE (at 6.25 ppm), MN (at 62.5 ppm), and CA (at 625 ppm) (*15*). These data on the cytogenetic effects of BD in rats versus mice are consistent with studies of the biotransformation of BD. Therefore, knowing how mouse strain differences in BD biotransformation will affect the potential genotoxicity of BD must await parallel mutation studies with B6C3F1 transgenic mice (*29*) and cytogenetic studies in MM.

The data presented in this chapter reveal that significant species differences in the V_{max} is observed for BD oxidation to BMO (see Table II). For example, B6C3F1 mouse liver microsomes displayed a capacity for BD oxidation exceeding that seen in either human or rat liver. This capacity was evidenced by a comparison of both the maximum velocity for the reaction, V_{max}, and V_{max}/K_M. Two putative detoxification enzymic reactions can occur with BMO, hydrolysis by epoxide hydrolase and conjugation with GSH by glutathione transferase. The results from our studies reveal that liver tissues from all species can detoxify BMO by both pathways (Tables III and IV). In general, human liver microsomes hydrolyzed BMO at greater rates than those of either rats or mice, as evidenced by the higher V_{max} and V_{max}/K_M. Although there was a considerable range of epoxide hydrolase activities in liver microsomes of the three human samples investigated (V_{max} = 9–58 nmol/mg protein/min), values for the human samples were at least twofold greater than the V_{max} for rats and mice. The value reported by Kreuzer et al. (*30*) for epoxide hydrolase-catalyzed hydrolysis of BMO by microsomes from a single human liver sample (V_{max} = 14 nmol/mg protein/min) falls within the range of values for V_{max} in human liver microsomes reported in this chapter. Values for V_{max}/K_M reported by Kreuzer et al. (*30*) for rodents and the one human liver sample were similar to the values reported in this chapter. For all species, apparent K_Ms for hydrolysis and GSH conjugation (not assessed in MM) were significantly greater than for BD oxidation and, in the case of mice, BMO oxidation.

Studies on the biotransformation of BD by MM liver microsomes indicate that MM enzymes catalyze the conversion of BD to BMO and hydrolysis of BMO. A comparison of the initial rates (V_{max}/K_M) of BD metabolism to BMO and BMO hydrolysis between MM and B6C3F1 mouse liver microsomes indicates that there are also mouse strain differences in BD metabolism. The ratio of V_{max}/K_M for BD to BMO and BMO hydrolysis in MM liver microsomes was 225:1, whereas the same ratio in B6C3F1 mouse liver microsomes is 360:1. The lack of mutagenicity of BD in the liver of the MM may be due to an ineffective concentration of BMO in the liver. Detailed parallel studies on the biotransformation of BD, in vitro and in vivo, in B6C3F1 and MM are required to establish the role of BD bioactivation and detoxification in mediating the genotoxic effects of BD.

The interaction of chemicals or their metabolites with DNA is a major factor in chemical carcinogenesis. One potential strategy for more accurately estimating human health risks from exposure to DNA-reactive chemicals is to develop research that will provide and improve understanding of the mechanisms of action of chemicals within an exposure → tissue dose → cancer response paradigm. The development of quantitative linkages between exposure and response, which are based on biologically plausible mechanisms of action at exposure levels that are likely to be encountered by people, will significantly improve the risk assessments for human exposures to DNA-reactive chemicals. This study was initiated to begin to establish in vivo linkages between exposure to BD, internal dose, and a biological response (mutation). The induction of mutations is a determinant of BD-induced carcinogenicity. Although these studies did not investigate the carcinogenicity of BD in MM, the induction of mutations in a transgene integrated into the genome of the MM represents an initial step toward establishing dose–response relationships for molecular events (mutation) that are part of the carcinogenic process. These relationships need to be established with endogenous genes that are known to be involved in the carcinogenic process (e.g., oncogenes and tumor suppressor genes). The in vitro metabolic constants from these studies can be incorporated into physiological models that can simulate in vivo behavior, and the models can be used to predict blood and tissue concentrations of BD and BMO. Experiments using whole animals can then be used to verify model predictions based on in vitro-derived rates. Ultimately, in vitro-derived rates for human tissue can then be used to predict BD and BMO concentrations in human tissues as a first step in estimating risk due to BD exposure.

Acknowledgments

The authors gratefully acknowledge the numerous valuable discussions with a number of our colleagues.

References

1. Harris, C. C. *Cancer Res.* **1991**, *51*, 5023s–5044s.
2. Weinstein, I. B. *Cancer Res.* **1988**, *48*, 4135–4143.
3. Shugaev, B. B. *Arch. Environ. Health* **1969**, *18*, 878–882.

4. Carpenter, C. P.; Shaffer, C. B.; Weil, C. S.; Smyth, H. R. *J. Ind. Hyg. Toxicol.* **1944**, *26*, 69–78.
5. Brunnemann, K. D.; Kagan, M. R.; Cox, J. E.; Hoffmann, D. *Carcinogenesis* **1990**, *11*, 1863–1868.
6. Pelz, N.; Dempster, A. M.; Shore, P. R. *J. Chrom. Sci.* **1990**, *28*, 230–235.
7. Environmental Protection Agency *Fed. Regist.* **1991**, *56*, 28548–28557.
8. Huff, J. E.; Melnick, R. L.; Solleveld, H. A.; Haseman, J. K.; Powers, M.; Miller, R. A. *Science (Washington, D.C.)* **1985**, *227*, 548–549.
9. Melnick, R. L.; Huff, J.; Chou, B. J.; Miller, R. A. *Cancer Res.* **1990**, *50*, 6592–6599.
10. Owen, P. E.; Glaister, J. R.; Gaunt, I. F.; Pullinger, D. H. *Am. Ind. Hyg. Assoc.* **1987**, *48*, 407–413.
11. Goodrow, T.; Reynolds, S.; Maronpot, R.; Anderson, M. *Cancer Res.* **1990**, *50*, 4818–4823.
12. de Meester, C.; Poncelet, F.; Roberfroid, M.; Mercier, M. *Toxicol. Lett.* **1980**, *6*, 125–130.
13. Poncelet, F.; de Meester, C.; Duverger van Bogaert, M.; Lambotte-Vandepaer, M.; Roberfroid, M.; Mercier, M. *Arch. Toxicol. Suppl.* **1980**, *4*, 63–66.
14. Duverger, M.; Lambotte, M.; Malvoisin, E.; de Meester, C.; Poncelet, F.; Mercier, M. *Toxicol. Eur. Res.* **1981**, *3*, 131–140.
15. Tice, R. R.; Boucher, R.; Luke, C. A.; Shelby, M. D. *Environ. Mutagen.* **1987**, *9*, 235–250.
16. Cunningham, M. J.; Choy, W. N.; Theall, A. G.; Rickard, L. B.; Vlachos, D. A.; Kinney, L. A.; Sarrif, A. M. *Mutagenesis* **1986**, *1*, 449–452.
17. DuBridge, R. B.; Calos, M. P. *Mutagenesis* **1988**, *3*, 1–9.
18. Gossen, J. A.; DeLeeuw, W. F. J.; Tan, C. H. T.; Zwarthoff, E. C.; Berends, F.; Lohman, P. H. M.; Knook, D. L.; Vijg, J. *Proc. Natl. Acad. Sci. U.S.A.* **1989**, *86*, 7971–7975.
19. Csanady, G. A.; Guengerich, F. P.; Bond, J. A. *Carcinogenesis* **1992**, *13*, 1143–1153.
20. Lowry, O. H.; Rosenbrough, N. F.; Farr, A. L.; Randall, R. J. *J. Biol. Chem.* **1951**, *193*, 265–275.
21. Omura, T.; Sato, R. *J. Biol. Chem.* **1964**, *239*, 2370–2378.
22. Hinners, R. G.; Burkart, J. K.; Punte, C. L. *Arch. Environ. Health* **1968**, *16*, 194–206.
23. Sambrook, J.; Fritsch, E. F.; Mainiatis, T. *Molecular Cloning: A Laboratory Manual*, 2nd ed.; Cold Spring Harbor Laboratory: New York, 1989.
24. Recio, L.; Osterman-Golkar, S.; Csanady, G. A.; Turner, M. J.; Myhr, B.; Moss, O.; Bond, J. *Toxicol. Appl. Pharmacol.* **1992**, *117*, 58–64.
25. Northrup, D. B. *Anal. Biochem.* **1983**, *132*, 457–461.
26. Bond, J. A.; Dahl, A. R.; Henderson, R. F.; Dutcher, J. S.; Mauderly, J. L.; Birnbaum, L. S. *Toxicol. Appl. Pharmacol.* **1986**, *84*, 617–627.
27. Kreiling, R.; Laib, R. J.; Filser, J. G.; Bolt, H. M. *Arch. Toxicol.* **1986**, *58*, 235–238.
28. Bolt, H. M.; Filser, J. G. *Arch. Toxicol.* **1984**, *55*, 213–218.
29. Kohler, S. E.; Provost, G. S.; Freck, A.; Kretz, P. L.; Bullock, W. O.; Sorge, J. A.; Putman, D. L.; Short, J. *Proc. Natl. Acad. Sci. U.S.A.* **1991**, *88*, 7958–7962.
30. Kreuzer, P. E.; Kessler, W.; Welter H. F.; Baur, C.; Filser, J. G. *Arch. Toxicol.* **1991**, *65*, 59–67.

RECEIVED for review September 3, 1992. ACCEPTED revised manuscript January 26, 1993.

12

Molecular Epidemiology of Acrylonitrile
Indicators of Health Risk by Worker Surveillance and Regiospecific Modification of Ha-*ras* Oncogene

John L. Wong[1], Bo Yuan[1], Peide Zhang[1], and Carlo H. Tamburro[2]

[1]Department of Chemistry, University of Louisville, Louisville, KY 40292
[2]Departments of Medicine and of Pharmacology and Toxicology, University of Louisville, Louisville, KY 40292

> *Prospective medical surveillance of chronically exposed chemical workers have shown no correlations between the rank-ordered estimates of acrylonitrile (AN) exposure and abnormal liver function, hepatotoxicity tests, or cancer development. Thus far, no dose–response effect of AN has been detected by medical screening of the workers in the surveillance program. We have used a molecular model of Ha-ras oncogene activation to evaluate the carcinogenic potential of acrylonitrile epoxide (ANO). The exogenous Ha-ras gene, in a superhelical plasmid or chromatin (active or quiescent), after treatment with a synthetic epoxide metabolite, has been found to be a useful indicator of carcinogenicity. At the cellular level, the epoxide-modified ras DNA is expressed in NIH3T3 cells to detect transforming activity. At the molecular level, the promutagenic lesions in plasmid and chromatin Ha-ras DNA are mapped by endonuclease S_1. In both comparisons, ANO and anti-benzo[a]pyrene-7,8-diol 9,10-epoxide (anti-BPDE) appear to be diametrically opposite. The convergence of human epidemiological and molecular data purport to indicate the relatively low health risk of AN as a carcinogen.*

ACRYLONITRILE (AN), CH_2=CH—CN, is an industrial vinyl monomer used in the manufacture of acrylic fibers, nitrile elastomers, and resins. AN is produced at a rate of ~1 billion kg/year in the United States and ~125,000 persons were estimated to be exposed in workplaces, according to the National Institute for Occupational Safety and Health (NIOSH) (*1*). AN is a volatile liquid, and it enters the body by inhalation or by skin absorption.

0065–2393/94/0241–0153$08.18/0
© 1994 American Chemical Society

Its acute toxicity and highly irritant effect on the respiratory tract, eyes, and skin are well-recognized; nausea, vomiting, vertigo, tremors, and unconsciousness can accompany excessive exposure (2). With long-term, chronic exposure, symptoms appear related predominantly to the autonomic nervous system, but its human carcinogenic potential is less certain (2). Epidemiological studies of exposure populations have yielded mixed evidence of increased cancer mortality. An earlier report (3) on several epidemiological studies of the rubber industry, which uses AN as well as many chemicals, identified excess mortalities from certain specific cancers (stomach, colon, and prostate in addition to neoplasms of the lymphatic and hematopoietic systems). Proportional mortality analysis at other small plants using AN revealed similar excesses for these cancers and some excess for lung (4), bladder, and central nervous system cancers (5). However, the most recent study (6) showed that AN is not associated with an increase in lung cancer or cancer of any type. Overall, the vinyl monomer AN exposure does not show the clear-cut evidence for carcinogenesis, as does vinyl chloride (VC), $CH_2=CH-Cl$, which is a well-documented human carcinogen (7).

We have been investigating occupational cancers, particularly those related to the vinyl monomers, since 1974 (7, 8). We have been working with local polymerization chemical plants in which synthetic rubber and plastics are produced, starting with AN and VC as the respective base chemicals. A rank-ordered index (ROI) system for estimating chemical exposure was initiated in response to the discovery of VC-induced hepatic angiosarcoma and liver injury. This approach was necessitated by the long latency period of cancer, the lack of prior environmental exposure measurement, the exposure to multiple chemicals in the work environment, and the need for a retrospective–prospective system for accurately estimating exposure to each chemical. Our occupational surveillance system includes two basic components: a complete work history with exposure rank orders for each major chemical in use and coded diagnoses covering medical tests and diseases. The ROI system's aim is to provide, for each employee, "exposure profiles or indices" for the estimation of health risk to each chemical within the surveillance program. The rank-ordered estimates of AN exposure have been correlated with the biochemical liver tests of the exposed workers. Studies to date (7) have found no dose–response relationship with AN and no indication that AN is like VC; impairment of liver functions and cancer development have followed increasing exposures to VC.

Nevertheless, in carcinogenesis bioassays, AN has been shown (2) to be a carcinogen in the rat, although this is not confirmed in any other species. The putative metabolite, acrylonitrile epoxide (ANO), was found (9) to be mutagenic in the Ames test and in human lymphoblast, in which AN itself is weakly genotoxic and requires metabolic activation (10). However, the ANO-DNA adducts remain elusive. When ANO was administered to rats, covalent binding to proteins occurred, but nucleic acid adducts could not be detected at the level of 0.3 alkylation per 1×10^6 bases (*11*). Therefore, a practical question in AN carcinogenesis is whether AN poses a significant health risk to workers in the normal work environment. Because extrapolation of animal tumor data to humans is debatable owing to dose and species differences, we have proposed a molecular model of oncogene activation to provide a contrast between established carcinogens and AN. Thus, we question whether ANO will modify a protooncogene to yield a transforming DNA for the induction of tumors. A causal relationship may exist between activation of the *ras* family of oncogenes and the development of human and animal tumors (*12*). In the case of hepatic angiosarcomas associated with occupational exposure to VC, DNA prepared from either frozen or paraffin-embedded tissue was examined for point mutations of c-*ras* genes (*13*). A mutation at codon 13 of the c-Ki-*ras* gene was detected in five of six tumors. It is therefore relevant to make use of Ha-*ras*-1 DNA in the chimeric plasmid pSV2neo-Ha-*ras* to study the genoactivity, mutational or transcriptional, of ANO and related carcinogens. In a nonfocus NIH3T3 transfection–transformation assay in which the epoxide-modified Ha-*ras* DNA is introduced into the cell chromosomes, we found that ANO showed negligible transforming potential, compared with the classical carcinogenic epoxide, *anti*-BPDE. It would seem that the vinyl chloride epoxide (VCO) would have made for a better comparison with ANO, except that VCO is known to rearrange rapidly to chloroacetaldehyde (CAA). Both VCO and CAA are reactive electrophilic species and can react with nucleic acid bases. The question of which is most important has been examined, for example, recently by Guengerich (*14*), but it is still unclear whether the c-Ki-*ras* mutation observed in the VC-induced tumors is a consequence of DNA adduct formed from VCO or CAA. Thus, we have chosen the classic and well-established epoxide carcinogen, *anti*-BPDE, to be the carcinogenic reference for ANO. This leads us to examine whether ANO is reactive toward Ha-*ras* DNA and whether different regiospecificities or extents of chemical promutagenic lesions may have been induced in Ha-*ras* DNA by the two epoxides. To this end, we have undertaken various modifications of Ha-*ras* DNA by ANO or *anti*-BPDE. First, the superhelical plasmid pSV2neo-Ha-*ras* was treated in vitro with the epoxides. Also, NIH3T3 cells transfected with the normal *ras* plasmid were briefly exposed to the epoxides. The in vivo experiment was designed to locate the major sites in chromosome-integrated Ha-*ras* DNA in transcriptionally active or inactive chromatin, where a DNA sequence, because of its conformational stress, may offer particular base positions as ready targets to the epoxide. Evidence for such position bias has come from many carcinogen–DNA studies (*15*). The alkylated bases with an altered hydrogen-bonding scheme may undergo unpairing of the duplex DNA and hence can be mapped by endonuclease S_1 according to the procedure described by Kohwi-Shigematsu and Nelson (*16*). This technique of mapping the major modification sites in Ha-*ras* DNA by the two epoxides ANO and *anti*-BPDE has shown regiospecific patterns with Ha-*ras* DNA of specific conformations, which appear to distinguish between their opposite transforming activity. Thus, a hypothesis of an association between *ras* gene activation by an epoxide and its human carcinogenicity can be tested in our present protocols of human surveillance and the molecular model of Ha-*ras* activation. The convergence of human and molecular studies may separate the agents at high risk of inducing human cancers from those at lower risk. Such classification or "proof of the negative" would have considerable practical interest. Our preliminary observation from both worker surveillance and the molecular model of Ha-*ras* activation purport to indicate the relatively low health risk of AN.

Experimental Methods

Rank-Order Estimates of Environmental Exposure to AN and Medical Screening. A cumulative-exposure rank system similar to that described by Smith et al. (*17*) was instituted and validated. The retrospective exposure indices were determined for each year and for each job category on the basis of payroll charge cards, starting with the plant's opening in 1940 and continuing until 1974. The prospective-exposure indices were begun in 1974 and updated annually. Exposure indices for each of the 22 chemicals [chemical code 01, acrylic acid; 02, acrylamides; 03, acrylonitrile; 04, acetylene; 05, acrylates; 06, bisphenol A; 07, butadiene; 08, caprylyl chloride; 09, chlorinated solvents; 10, chloroethyl vinyl ether; 11, diethyl maleate; 12, mercuric chloride; 13, methanol; 14, phenol; 15, toluene; 16, vinyl chloride; 17, vinylidene chloride; 18, vinyl acetate; 19, poly(vinyl chloride) (PVC) dust; 20, catalysts; 21, styrene; and 22, hexane] were created for each employee on the basis of the employee's yearly work history and rank-ordered job-exposure categories. Chemical exposure was rated by a seven-category ordered ranking, by which levels of exposure were rated from 0 through 6. Ratings were assigned by consensus of a panel of three to five chemical-exposure judges, who were mostly production foremen, chemical engineers with experience in the given work area, or individuals who had worked during the year being evaluated. An example of a chemical exposure record for an authentic individual worker is partially shown in Table I.

Analytical monitoring of AN was subsequently carried out. The monitoring was done by using an activated charcoal sampling tube fastened to the worker's shoulder and a Bendix pump fastened to the belt was worn during the entire work shift. The air flow rate was 200 mL/min, and the adsorbed AN was later desorbed by carbon disulfide. This method is similar to the NIOSH method S156 (*18*), and has been validated over a concentration range of 7.5–70.0 mg/M^3 (3.6–33.3 ppm), coefficient of variation 0.073. The simple modification of method S156 by using a desorbing solvent of 2% V/V acetone in carbon disulfide (*19*) allows achieving sensitivities of 1.1 mg/

Table I. Detailed Work and Exposure History

Year	Work History			Exposure Rank of Chemical[a]
	Building	Job	No. Months	
1944	000	576	6	1
1945	000	576	5	1
1945	111	194	7	1
1946	111	194	12	1
1947	111	194	12	3
1948	111	194	8	3
1948	121	192	4	1
1974	Terminated			

[a] Acrylonitrile was the chemical.

M^3 or 0.5 ppm based on air sample volume of 15 L. Analysis was performed on a gas chromatograph on 10-ft × 1/8-in. SP 1000 operated isothermally at 100 °C, calibrated over the range of 20 to 100 μg/tube.

In this worker cohort, medical data regarding hepatotoxicity were annually obtained to identify injuries that may result from chemical exposure. Over 80% of the work force had been screened for hepatotoxicity by the serum tests alanine aminotranferase (ALT), aspartate animotransferase (AST), and alkaline phosphatase (AP) as well as liver functional studies by serum albumin, prothrombin, bile acids, and indocyanine green (ICG) clearance (20). The liver test results were evaluated for the frequency of abnormalities seen among the chemical workers with various exposure rankings to the suspect chemical AN.

Epoxide Modification of Ha-ras Plasmid DNA. The molecular cloning of pSV2neo-Ha-ras was done by ligating a BamHI fragment of human Ha-ras-1 [from American Type Culture Collection (ATCC) 41,001] to the pSV2neo vector (ATCC 37,149); a clone of HB 101 transformant with the correct order of ligation, as shown by gel electrophoresis, was selected. Chemicals were purchased from Aldrich, unless otherwise noted. The preparation of ANO was done by fractional distillation of a hypochlorite oxidation mixture of AN (21) and its purity verified by ^{1}H and ^{13}C NMR. To 65 μL of medium made up of 1:1 tetrahydrofuran and 20 mM ethylenediaminetetraacetate (EDTA) buffer pH 8.5 (TE) containing 10 μg of pSV2neo-Ha-ras was added 1% ANO in tetrahydrofuran in microliter quantities to obtain 40 μM. The reaction of plasmid DNA, 10 μg, with *anti*-BPDE (purchased from Midwest Research) in 0.1 mL of 20 mM TE (pH 7.6) at 20 μM was performed as before (21). All the reaction mixtures were incubated at 37 °C for 2.5 h, cooled, extracted with 3 × 50 μL of ether, followed by 0.1 volume of 3 M NaOAc (pH 7.2) and 2.5 volume of absolute ethanol to precipitate DNA.

Epoxide Modification of Chromatin DNA in NIH3T3 Ha-*ras* Transfectants.

NIH3T3 cells (1×10^6 cells enough for 10 subsequent experiments) were transfected with 20 μg of pSV2neo-Ha-*ras* DNA, and clones were selected with G418 (400 μg/mL) (21). Half of the transfectants were stimulated by exposure to 12-O-tetradecanoylphorbol-13-acetate (TPA, Sigma Chemical Company) at 50 ng/mL in the complete medium. After 12 h, the culture media were replaced in both the TPA-treated and untreated NIH3T3 transfectants with a buffer containing 0.15 M NaCl, 1 mM EDTA, 30 mM Tris-HCl, and 2 mM $ZnCl_2$. The cultures in triplicates were treated with either the ANO solution (80 μM total added in three aliquots) or with *anti*-BPDE (20 μM) at 37 °C for 2 h. The remaining cultures were given solvents only as negative controls. The genomic DNA was isolated as described previously (21).

Mapping of Epoxide Modified Ha-*ras* Plasmid by Endonuclease S_1.

To a solution of the modified Ha-*ras* plasmid DNA, 10 μg in 5 μL of water was added 2 μL of 5 × restrictive enzyme (RE) buffer (Bethesda Research Laboratories [BRL] #3), 2 μL of water, and 1 μL of EcoRI (BRL, 5 units/μL), and the mixture was incubated at 37 °C for 4 h. The DNA was precipitated as already described. Addition of 5 μL of water, 2.5 μL of 0.2 mM of deoxythymidine triphosphate, deoxycytidine triphosphate, deoxyguanosine triphosphate (BRL), 1.25 μL of 0.4 mM of deoxyadenosine triphosphate (dATP)-7-biotin (BRL), and 1.0 μL of DNA polymerase I (klenow fragment, 5 units/μL, Promega) followed; the mixture was incubated at 15 °C for 30 min, and biotin labeling was stopped by adding 1.25 μL of 0.5 M EDTA (pH 7.8). After adding 0.1 volume of 3 M NaOAc (pH 7.2) and 2.5 volume of absolute ethanol, the solution was stored at -20 °C overnight, the precipitated DNA spun in a microfuge for 10 min, and the dry pellet redissolved in 16 μL of water. The unwanted 5' end label was removed by using a solution made of 2 μL of 10 × RE buffer (BRL #5) and 2 μL of XmnI (BRL, 5 units/μL), and the mixture was incubated at 37 °C for 4 h. The DNA was precipitated in absolute ethanol and the pellet redissolved in 4 μL of water. To the solution were added 5 μL of 2 × S_1 nuclease buffer (30 mM NaOAc at pH 4.6, 50 mM NaCl, 1 mM $Zn(OAc)_2$, 5% glycerol, and 500 μg/mL of salmon sperm DNA) and 1 μL of endonuclease S_1 (BRL, 5 units/μL), whereupon it was incubated at 37 °C for 10 min and the reaction was stopped by adding 16 μL of TE at pH 7.6. The DNA was analyzed by electrophoresis at 45 V in Tris-acetic acid-EDTA buffer on 1% agarose gel for 5 h. The nick-translated, biotin-7-dATP-labeled lamda DNA Hind III digest (Sigma) was used as the molecular weight markers. The DNA was transferred to nitrocellulose filter paper from the agarose gel by capillary action in 20 × sodium chloride–sodium citrate (SSC) stock solution overnight and the filter baked at 80 °C for 2 h under vacuum; this was followed by colorimetric detection to visualize the bands.

Mapping of Epoxide Modified Ha-*ras* Chromatin DNA by Endonuclease S_1. The purified cellular DNA pellet, 10 μg, was dissolved in 10 mM Tris-HCl (pH 7.2) and 1 mM EDTA. Digestion of the genomic DNA was carried out with BamHI, followed by S_1 nuclease (BRL) at 0.5–1 unit/μg of cellular DNA in the S_1 buffer containing 3 mM NaOAc, 100 mM NaCl, and 2 mM $ZnCl_2$, at pH 4.5 and 37 °C for 10–15 min. The reaction was terminated by adjusting the pH to 7.0 with 2 M Tris-HCl (pH 8.0), at which point S_1 nuclease became inactive. The DNA was precipitated with ethanol, dissolved in TE buffer, separated on 1% agarose gel, and transferred to nitrocellulose (NC) paper in 20 × SSC. The filter was hybridized with a nick-translated, biotin-labeled BamHI-ClaI (870 base pair) probe of Ha-*ras* DNA obtained from pbc-N1 (ATCC 41,001) for Southern blot.

Results and Discussion

Retrospective Validation of Rank-Order Estimates of AN Exposure. The epidemiological identification of occupational carcinogens is complicated by several problems, including worker mobility between jobs, variations over time of chemicals and processes used, and long latency between exposure and discovery of the disease. In addressing these problems, we have obtained qualitative exposure estimates. Table I illustrates the format to record the combined work history and exposure rank of an individual worker for each of the 22 chemicals used in the plant, including AN as shown. In this case, in part A, the individual began working in 1944 and worked for 6 months in job #000 576 (area and job three-digit code). He worked an additional 5 months on this job in 1945. He completed the last 7 months of 1945 in another job #111 194. Part B of Table I identifies the exposure ratings assigned to AN for each job number for each year. For example, in the first row the exposure rating for AN assigned in 1944 to job #000 576 is indicated. In 1945, this remained unchanged. The third row records the exposure rating of job #111 194 during 1945. The other 21 chemicals were similarly rated, although their rankings are not shown here. The information in parts A and B may be combined to calculate the cumulative exposure rank months (CERM) for a certain chemical. In this example, the individual in 1944 would have had six CERMs for having worked 6 months at an exposure rank of 1 during that year. During 1945, this individual would have had a CERM for AN of $(5 \times 1) + (7 \times 1)$, or 12. In contrast, in 1948 he worked 8 months at a job with rank 3 and 4 months at a job with rank 1, giving him a total of 28 CERMs of AN. The initial assumption was made that the CERMs could be used as a rank-ordered statistic. Their purpose is to define and identify groups of workers who have a common occupational exposure, having performed the same or similar types of jobs. The predictive value of CERM was empirically validated for the association between VC and hepatic angiosarcoma, as well as VC-induced hepatic injury and fibrosis (22).

In 1978 analytical monitoring for AN was begun in response to OSHA requirement. The rating of exposures obtained from the chemical-exposure judges was continued but determined independent of the analytical determination of AN. Between 1975 and 1979, the synthetic rubber manufacturing process remained essentially unchanged. Analytical monitoring for AN was begun in late 1977, making 1975 and 1976 the last 2 complete years during which a rank-ordered estimate was obtained without quantitative monitoring data for AN being available. The years 1978 and 1979 were the first 2 complete years in which quantitative monitoring data for each job were available for AN. The opportunity, therefore, was present for a comparison of the first 2 years' analytically determined exposure ranking of AN with the last 2 years' qualitative estimate. The sample size for each job classification varied from 5 to 135, the average being 10, with 50% of the jobs having 14 or more samples. Over the 1-year period, 80% of the 60 job classifications had 3 or more days of AN monitoring, 70% having 5 or more days of monitoring. Only six jobs (10%) were monitored for a single day. A comparison of job-specific AN exposures determined analytically with job-exposure ranks in previous years is shown in Figure 1. The mean time-weighted average (TWA) AN levels in parts per million appear to correlate with the exposure rank orders given to the specific jobs (Figure 1A). For the 1978 analysis versus 1975 rank order, jobs rank-ordered at 1 had mean exposure levels below 1 ppm, whereas all those with rank orders of 4 and 5 were above 1 ppm. For jobs rank-ordered at 2, 80% had mean AN levels of 1 ppm, whereas in jobs at ranks of 3, 80% had mean AN levels above 1 ppm. For the 1979 analysis versus 1976 rank order, all jobs rank-ordered at 1 and 2 had mean AN levels of less than 1 ppm; at rank order 3, 30% were less than 1 ppm and 70% were greater than 1 ppm; all jobs rank-ordered at 4 and 5 were greater than 1 ppm. Data of both comparison groups show low rank-ordered jobs with mean AN exposure at less than 0.6 ppm level, whereas higher rank-ordered jobs (3, 4, and 5) are at the level of 1–3 ppm.

Individual maximum TWA exposure levels to AN further demonstrate the correlation between rank order, given a specific job, and the job's individual analytically determined exposure. Figure 1B shows this comparison for the same periods. For the 1978 analysis versus 1975 rank order, 90% of jobs rank-ordered at 1 and 2 were associated with maximum exposure levels below 3 ppm, averaging approximately 1 ppm, but no job had maximum exposures of more than 10 ppm. All jobs rank-ordered at 3, 4, and 5 had maximum AN levels above 1 ppm, some jobs having maximum exposures of up to 90 ppm. Similar levels were found for exposures determined in the 1979 versus 1976 comparison. In both study periods, 90% of jobs rank-ordered at 4 and 5 had maximum AN levels above 13 ppm; 15% had maximum AN levels above 20 ppm. The rank-ordering relative to exposure measure-

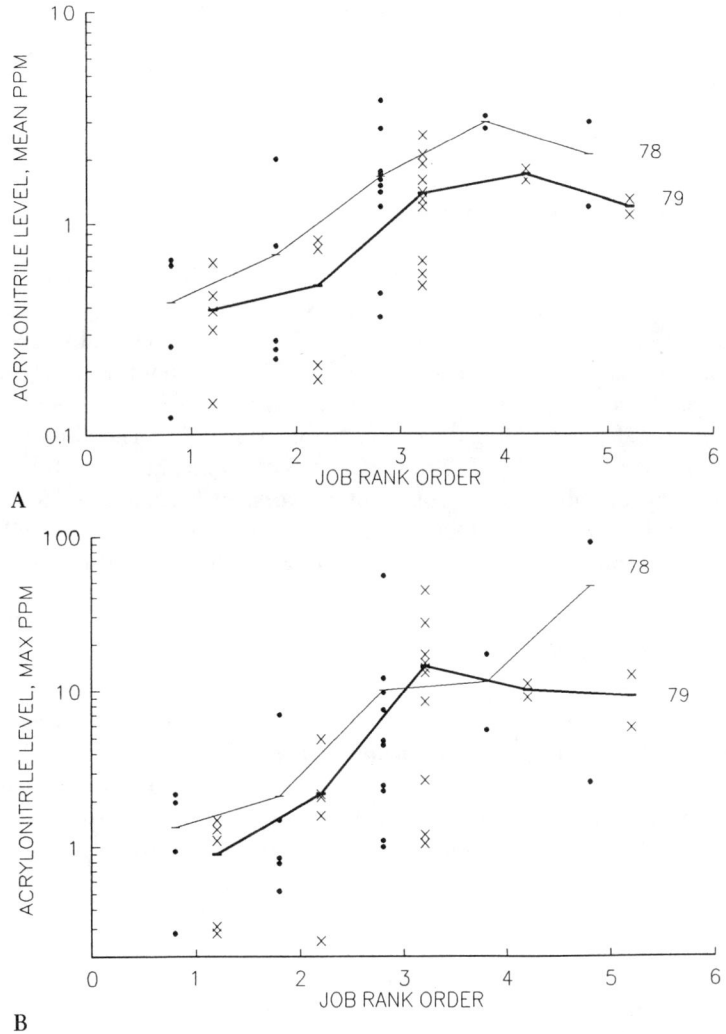

Figure 1. Comparison of job-specific AN exposure levels (ppm) determined by personal monitors with estimates of job rank-orders for each job; • indicates analytical AN ppm (1978) versus job rank (1975), and x indicates analytical AN ppm (1979) versus job rank (1976). A, mean AN exposure values for comparison. B, Maximum AN exposure values for comparison.

ments are consistent. The mean levels, in contrast to maximum levels, show that this proportionate relationship levels off in the higher-rank-ordered jobs but is maintained with the lower-rank-ordered jobs. This effect is due to environmental controls to maintain an AN work level of OSHA standard. Therefore, higher-ranked jobs (those with more intimate chemical contact) are also those with greatest environmental control. These measures keep the mean AN levels below 5 ppm (Figure 1A). When excursion occurs and leads to maximum levels, one sees that the AN level continues a linear correlation for 1978, even with higher job ranking (Figure 1B). The 1979 analysis, however, showed lower maximum levels than 1978, the first year of environmental monitoring. The leveling-off in the 1979 curve reflects more conscientious efforts by the industry and its workers to be in compliance with the OSHA guidelines instituted 1 year ago.

Correlation of Hepatotoxicity Tests with Cumulative AN Exposure. The OSHA-regulated levels of AN allow 2 ppm (8 h TWA), with 1 ppm as action level and 10 ppm ceiling for 15 min (*1*). Some jobs in Figures 1 and 2 have exceeded these levels. It will be of practical interest to correlate the hepatotoxicity test results with the estimated CERMs for AN for the same period. The screening data were obtained in 1979. The abnormal values, defined as values above the upper limits of normal range for the control group, were identified in this cohort of workers. In Figures 2A, 2B, and 2C are plotted frequency as percentage of abnormal tests versus AN CERM, at 100-unit increments up to 1600. This frequency is shown for three representative screening tests: ALT, AP, and ICG clearance, respectively. The first two are federally required; ALT reflects high sensitivity and has the highest specificity for hepatic injury. The ICG clearance, however, provides the best combination of positive predictive value and sum of specificity and sensitivity (*20*). ALT may be a substitute for the more complicated ICG clearance technique, if adequate medical facilities are not accessible. The first two plots have only random distribution and share no relationship between abnormal ALT or AP with cumulative AN exposures. For the ICG plot in Figure 2C, the percentage of abnormal tests versus VC CERM is also given. The fairly linear dose–response relationship as seen in the VC plot is not shown by the AN plot. The AN plot appears to show some increase in abnormal ALT or AP with high exposures, but the response is greatly dispersed. Indeed, a more detailed study of the workers with abnormal screening tests revealed that most of this relationship was due to the concomitant exposure to VC. This diagnosis was further supported by examination of the specific components of the hepatic tissue damage (*22*). Medical data continue to be collected from this cohort. We have not found any significant disease–AN cumulative chronic-exposure relationship. However, these negative findings often raise questions about the limitation of the CERM approach and the need for more direct biomarkers of biochemical events resulting from exposure to AN. This approach will be necessary for molecular epidemiology. Therefore, we have made use of a molecular model of Ha-*ras* oncogene DNA activation to compare the direct action of ANO with the established carcinogen *anti*-BPDE. This model study may form the basis for detecting activated *ras* oncogene from peripheral blood or from other biopsy tissues obtained from the AN-exposed workers.

Ha-*ras* DNA in Transfection–Transformation Assay. Activated transforming genes (oncogenes) have been found in a number of human tumors by use of assays in which transformed foci result from transfection of tumor DNA into NIH3T3 cells. One striking fact that has emerged from screening transfecting DNA with the NIH3T3 transformation assay is that, for both human and rodent tumor DNA, the transforming genes are virtually all related to the *ras* oncogene family (*12*). This high frequency of *ras* mutation observed in the NIH3T3 assay may simply reflect the susceptibility of this assay to transformation by an altered *ras*-dependent signal-transduction pathway. However, the *ras* oncogenes are among the best-studied human oncogenes because of their frequent detection in primary tumors and their apparently central role in human cancer. A review by Spandidos (*23*) has shown the involvement of the *ras* gene family in early, secondary, and late stages of carcinogenesis, as shown further on. A mutated Ha-*ras*, for example the EJ-*ras* gene isolated from a human bladder carcinoma cell line, does not require a cooperating gene to trigger malignant transformation. Thus, Ha-*ras* may be used as a critical oncogene model for transgenic study of ANO carcinogenesis at the molecular level. In comparison, the commonly used eukaryotic assays of carcinogens with endogenous genes like adenine phosphoribosyltransferase (aprt), dihydrofolate reductase (dhfr), and hypoxanthine-guanine phosphoribosyltransferase (hprt), provide less direct information on the carcinogenesis pathway because they are not oncogenes (*24*).

In our study of ANO, the transforming potential of *ras* DNA, chemically modified by ANO, is tested in a NIH3T3 transfection–transformation assay. The transfectants are subjected to triple selections: G418 resistance, low serum growth, and limit dilutions. The end points are scored by cell growth kinetics and by monolayer saturation density, which provide a quantitative characterization of the transformants, thereby obviating the capricious nature of counting foci for transforming potential. The advantage of this chemical approach is that it removes the uncertainty associated with the action of a carcinogen on DNA inside a cell. This approach points to the initiation of tumorigenesis by the modified Ha-*ras* DNA, because it is the only macromolecule present in the chemical reaction, which is followed by transfecting the modified DNA into NIH3T3 cells. Thus, the epoxide modification of the Ha-*ras* may occur without the complications arising from metabolic activation and detoxification, cytotoxicity, immune suppression, and induced mutation of an in vivo study. When transfected in NIH3T3 cells, the direct involvement of Ha-*ras* oncogene activation by the epoxide in the transformation will be obvious. The more common practice of testing a chemical in vivo and isolating the resultant tumor DNA for the NIH3T3 focus assay provides less direct information on its carcinogenic potential.

The chemical-bioassay approach makes use of a new chimeric plasmid, pSV2neo-Ha-*ras*. To illustrate its transforming potential, we have started with ligating the EJ-*ras* tumor DNA to the vector pSV2neo. Two types of

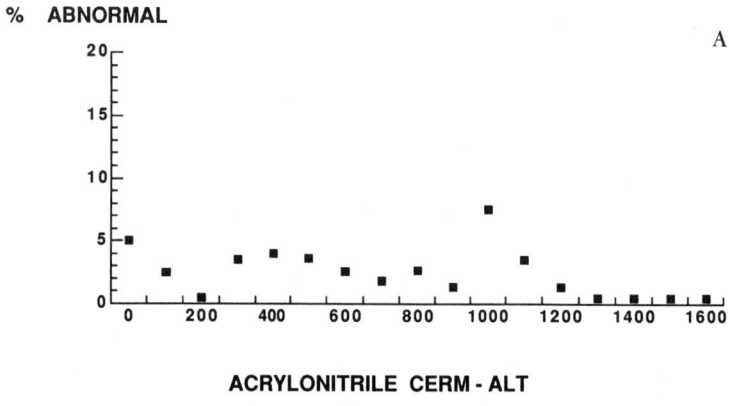

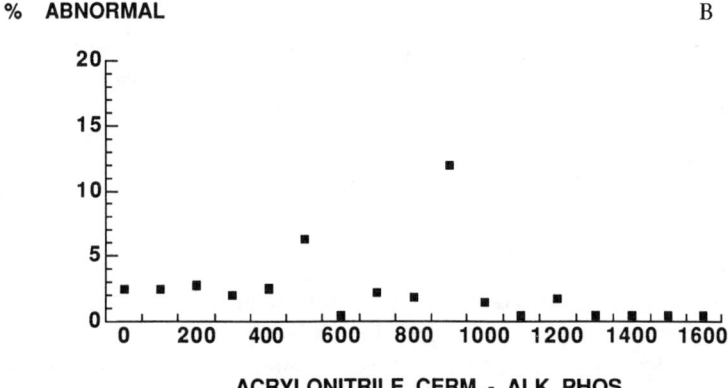

Figure 2. Biochemical liver tests of workers exposed to AN; screening results higher than the normal range of controls expressed as % abnormal in a group of workers with the same AN–CERM: A, ALT, B, AP.

HB101 transformants were isolated for both directions of ligation. One showed the 6.6-kb EJ-*ras* BamHI fragment inserted in the opposite direction (not promoted by SV40 promoter), as evidenced by restrictive analysis of its DNA with EcoRI/KpnI. In Figure 3B lane 2, two bands at 9.4 and 2.8 kb were found. The other transformant revealed two bands at 6.8 and 5.4 kb in the same restrictive analysis (lane 4), and also one and two bands, respectively, in lane 2 (EcoRI digest) and lane 5 (BamHI digest), thus verifying the proper insert (promoted by SV40). Furthermore, only the latter plasmid construct was able to induce focus formation in the NIH3T3 transfectant (see Figure 3C). Likewise, when the 6.4-kb BamHI fragment of c-Ha-*ras*-1 protooncogene was ligated to the pSV2neo vector, the desirable HB101 transformant was selected on the basis of the foregoing restrictive analysis. Figure 4 lane 5 showed the two bands at 6.8 and 5.2 kb from EcoRI/KpnI; lane 4 showed the EcoRI-linearized form at 12 kb, and lane 6 the two

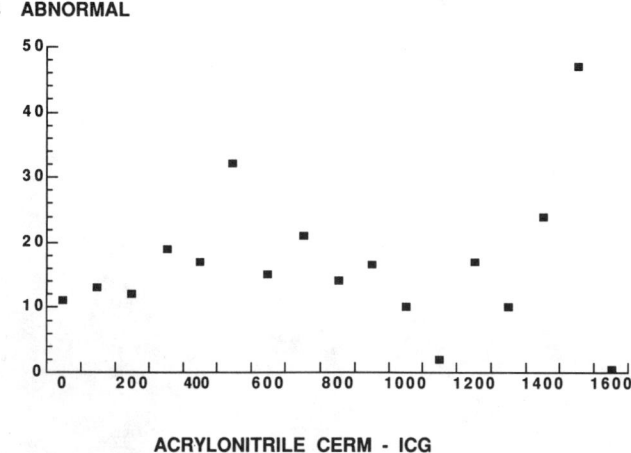

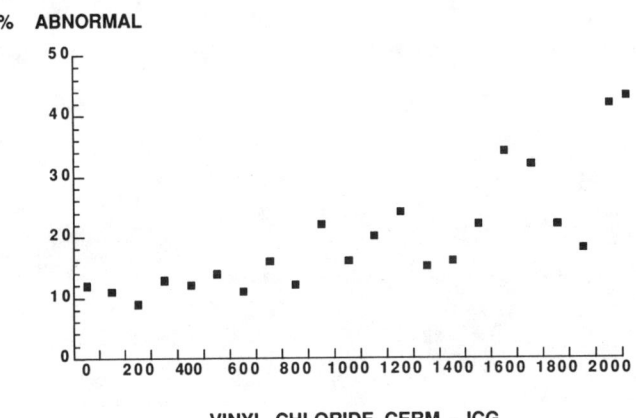

Figure 2. Part C. ICG clearance at 0.5 mg/kg body weight for AN and vinyl chloride cream.

BamHI fragments at 6.4 and 5.6 kb for the Ha-*ras* DNA and the vector pSV2neo, respectively. The correctly ligated plasmid, pSV2neo-Ha-*ras*, was cloned to obtain enough DNA for epoxide modification and the NIH3T3 assay. When the normal Ha-*ras* was treated with *anti*-BPDE or *N*-methyl-*N*-nitrosourea, the modified DNA was found positive in transforming NIH3T3 cells. Although ANO-modified Ha-*ras* gave rise to two G418[R] clones, both were unambiguously scored negative because of their normal growth rate and monolayer density similar to the negative controls (*21*).

Furthermore, *anti*-BPDE caused mutation at codon 12 of the *ras* oncogene as shown by Southern blot analysis, but ANO did not. These differences may exist because of different regiospecific alkylations of the DNA or different repairs of the modified DNA. We have addressed first the question

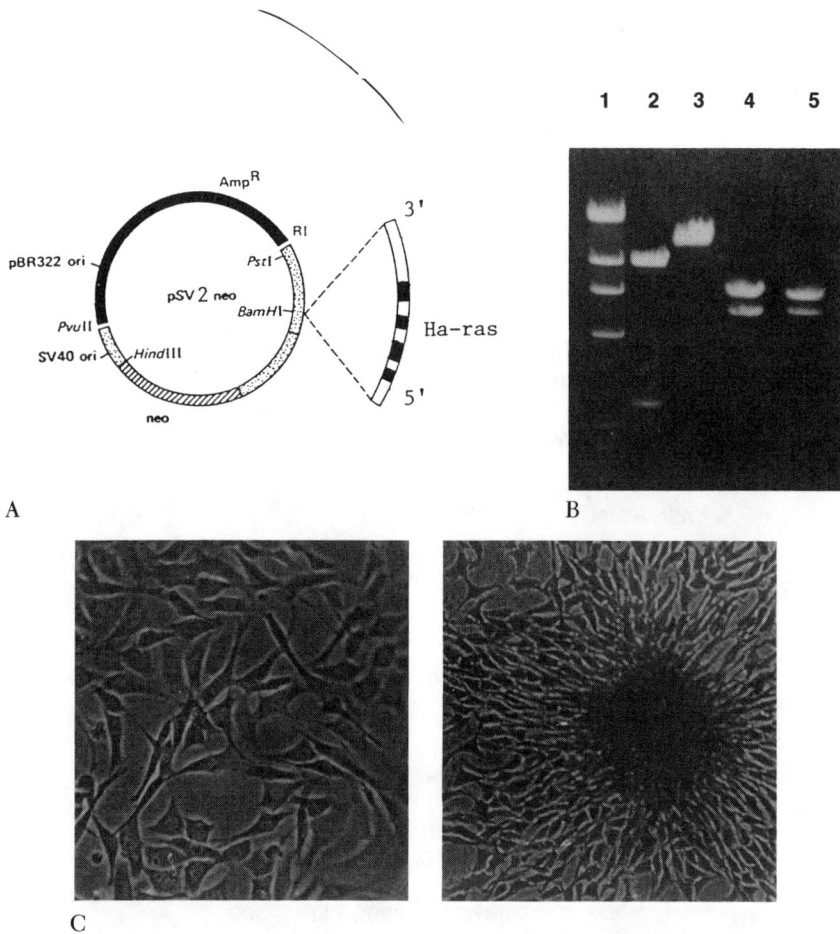

Figure 3. Construct, restrictive analysis, and transforming potential of NIH3T3 transfectant of pSV2neo-EJ-ras: A, Construct from pSV2neo (5.6 kb) + EJ-ras (6.6 kb), two directions of ligation, the correct one (with SV40 promoter) is selectable, B, agarose gel (0.8%) electrophoresis: lane 1 molecular weight markers, Hind III digest of λ-DNA: 23.1, 9.4, 6.6, 4.4, 2.3, and 2.0 kb; lane 2 digest of opposite direction plasmid pSV2neo-EJ-ras (not promoted by SV40 promoter) with KpnI followed by EcoRI; lane 3 EcoRI digest of the correct plasmid pSV2neo-EJ-ras (promoted by SV40); lane 4 restrictive analysis of plasmid in lane 3 using the same EcoRI/KpnI treatment as in lane 2; Lane 5 BamHI digest of pSV2neo-EJ-ras used in lane 3. C, Focus formation of NIH3T3 transfected with pSV2neo-EJ-ras having the correct ligation (lane 4 in B); controls (pSV2neo, pSV2neo-Ha-ras) did not give any focus or abnormal saturation density in confluent growth of NIH3T3 transfectants.

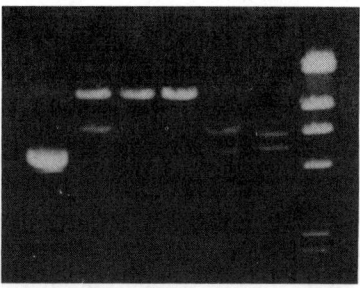

Figure 4. Restrictive analysis and endonuclease S_1 treatment of pSV2neo-Ha-ras: lane 1 pSV2neo-Ha-ras superhelical form dominant; lane 2 digest of plasmid with EcoRI and endonuclease S_1 (1 unit/μg, 37°C, 10 min); lane 3 has same treatment as in lane 2, except S_1 used at 0.5 unit/μg DNA; lane 4 digest of EcoRI only; lane 5 digest of plasmid with KpnI followed by EcoRI; lane 6 digest of BamHI; and lane 7 λ-DNA Hind III molecular-weight markers.

of whether ANO is capable of modifying Ha-*ras* DNA and whether the sites of alkylation of Ha-*ras* DNA are different for ANO and the established carcinogen *anti*-BPDE.

Chemical Modification of Ha-*ras* DNA by Epoxides. On the basis of molecular structure, the reactivities of ANO and *anti*-BPDE toward the DNA nucleophiles are shown in Figure 5. Stereospecific S_N1 reaction at the benzylic carbocation position of *anti*-BPDE with the primary target exocyclic amino group of guanine was reported by Jeffrey (25). However, similar

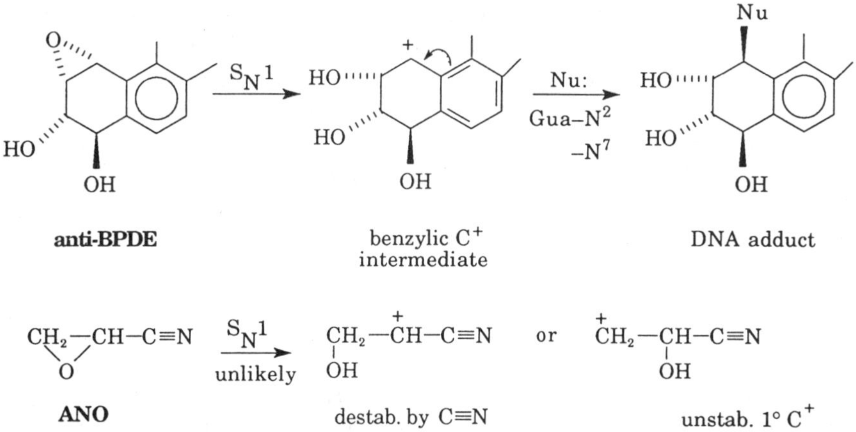

Figure 5. Comparison of heterolytic ring-opening pathways of anti-BPDE and ANO by DNA nucleophiles.

S_N1 pathways for ANO would generate unstable carbocation intermediates, and hence ANO would be less favored to react with DNA in the same manner as *anti*-BPDE. In sorting out their different chemical reactivities, we conducted in vitro reaction of ANO with plasmid Ha-*ras* DNA, the positive control being *anti*-BPDE, and the negative control the solvent-treated plasmid (mock reaction). The strategy to map the chemical modification sites in pSV2neo-Ha-*ras* plasmid DNA is based on the use of S_1 endonuclease. This mapping technique is predicated on the epoxide-modified base, on becoming unpaired, offering a cleavage site for the endonuclease S_1 (*16, 26*). The same site in the original DNA would be double-stranded and would not be so digested. Figure 4 shows the electrophoretic analysis of pSV2neo-Ha-*ras*. Lane 1 revealed the dominant superhelical DNA of the untreated plasmid, as opposed to the EcoRI-linearized form in lane 4. The combination of EcoRI and endonuclease S_1 did not generate any new band (see lane 3) when the S_1 was used under the same conditions of S_1-mapping of the epoxide-modified plasmid (0.5 unit/µg DNA, 37 °C, 10 min). However, at double the S_1 concentration, a new band at ~7 kb was observed in lane 2. Thus, the endonuclease S_1 sites so determined reflect only the epoxide-induced unpairing of the original base pairs, representing significant damage to the Ha-*ras* DNA. The S_1 sites are detected without the intervention of DNA repair or cell selection for a mutation that imparts growth advantage such as in codon 12 or 61. The technique for mapping modification is shown in Figure 6. From the Southern blot analysis (Figure 7), ANO generated four bands in the region of 2.0–6.6 kb, whereas *anti*-BPDE gave rise to six bands in the same region plus a few smaller fragments. These modifications were detected without the sensitivity of a hybridization probe; hence the 3'-end biotin labeling revealed only the prominent S_1 cleavage sites. The reactive sites are more likely to occur on the dominant superhelical circular form of pSV2neo-Ha-*ras*, as indicated by comparing lane 1 with lane 4 in Figure 4. One possibility is that the hot spots in the superhelical *ras* plasmid may be common to both epoxides. A precedent was reported by Kohwi-Shigematsu et al. (*26*) in the case of a supercoiled plasmid DNA harboring either inverted repeats or poly(dG)–poly(dC) sequences. Two different types of chemical carcinogens—*N*-acetoxy-2-acetylaminofluorene, which bonds to the C-8 position of guanine, and chloroacetaldehyde, which forms a cyclic etheno derivative with the H_2N—C=N— residue of DNA bases—appear to have modified the same specific DNA sites. Presumably, both carcinogens reacted with the bases, which were unpaired non-B DNA structures adopted by certain DNA sequences. However, the two epoxides reported herein behaved toward the Ha-*ras* DNA in different manners. The four bands produced by the in vitro reaction of ANO with Ha-*ras* DNA clearly show that ANO is DNA-reactive but has a regiospecificity unlike that of *anti*-BPDE. Although this gel analysis was not designed to have the resolution for sequence analysis, it is noteworthy that the band at 5 kb, which is very pro-

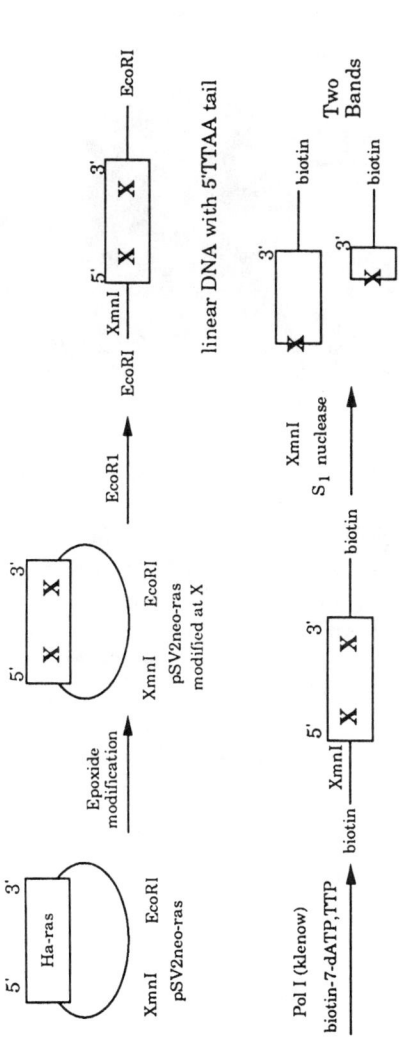

Figure 6. Strategy of endonuclease S_1 mapping of epoxide-modified sites in Ha-ras DNA contained in pSV2neo-Ha-ras plasmid.

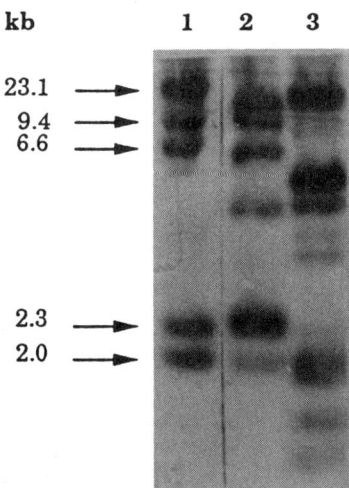

Figure 7. Endonuclease S_1 analysis of epoxide-modified pSV2neo-Ha-ras plasmid. Southern blot lane 1 molecular-weight markers, Hind III digest of λ-DNA, lane 2 ANO-modified, and lane 3 anti-BPDE–modified.

nounced in the *anti*-BPDE lane, is missing from that of the ANO as shown in Figure 7. This band corresponds approximately to 2.2 kb on the Ha-*ras* gene map, which puts it in the vicinity of codon (2110 bp). The latter codon has been found susceptible to modification by *anti*-BPDE. Vousden et al. (27) reported that a transforming oncogene resulting from in vitro modification of a Ha-*ras* plasmid with *anti*-BPDE involved mutation in the 61st codon (70% of transformants) and the 12th codon (30%). Although these mutations may be interpreted as selections through clonal expansion in tumors induced by BPDE, the lack of modification by ANO in the 61st codon area in the present *ras* DNA reaction appears to be consistent with the lack of transforming potential of the ANO-treated pSV2neo-Ha-*ras*.

Treatment of Chromatin Ha-*ras* DNA with ANO. The in vivo epoxide–chromatin DNA results, as shown in Figure 8, are clear-cut. The S_1 nuclease mapping technique did not reveal any modification site in the Ha-*ras* DNA isolated from the NIH3T3 transfectants after a brief exposure to ANO. Both the quiescent cells and the cells stimulated by the phorbol ester TPA showed only the 6.4-kb BamHI fragment in the isolated genomic DNA just like the BamHI fragment of the negative control cells (pSV2neo-Ha-*ras* transfectants exposed to solvent only). The quiescent cells exposed to *anti*-BPDE also gave the 6.4-kb *ras* fragment only. An additional major band of 6.2 kb, about 0.2 kb from the 5' end of the 6.4 kb BamHI Ha-*ras* fragment, was observed from the *anti*-BPDE treatment of the transcriptionally active, chromatin *ras* DNA. Regulatory sequences have been reported

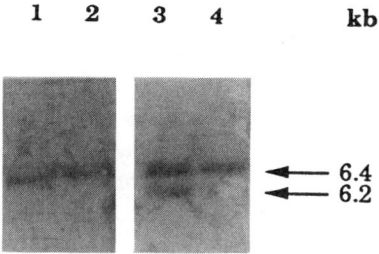

Figure 8. Endonuclease S_1 analysis of chromatin Ha-ras DNA isolated from NIH3T3 Ha-ras transfectants, Southern-blotted and probed with a biotin-labeled, nick-translated BamHI-ClaI 870 bp fragment of Ha-ras DNA from pbc-N1: lanes 1 and 2, ANO-exposed; lanes 3 and 4, anti-BPDE–exposed; lanes 1 and 3 for cells pretreated with TPA.

by Ishii et al. (28) in the far 5' region of the c-Ha-*ras*-1 gene. Pertinent to the present epoxide binding is that a promoter (nucleotides 117–536) was found to express chloramphenicol acetyltransferase (CAT) activity similar to that found when the long terminal repeat promoter of the Rous sarcoma virus was used to control CAT. Spandidos et al. (29) also reported a phorbol ester–responsive Ha-*ras*-1 gene promoter. They found that rat cell transfectants that carried CAT plasmids linked to the promoters in the SstI fragment of nucleotides 211–1063 of normal Ha-*ras*-1, on exposure to 50 ng TPA/mL for 12 h, showed a 15-fold increase in CAT expression over the untreated cells. The NIH3T3 Ha-*ras* transfectants treated with TPA and *anti*-BPDE possibly became transcriptionally active as well as reactive in the 5'-regulatory region. Thus, an abnormally high level of the *ras* protein p21, with or without other changes due to point mutations in the coding region, may result, thereby transforming the NIH3T3 transfectants. However, the lack of endonuclease S_1 sites in active chromatin *ras* DNA treated with ANO contrasts with the behavior of *anti*-BPDE. This new observation is an extension of the inactivity of ANO-modified *ras* in a nonfocus NIH3T3 transfection–transformation assay, which we reported earlier (21). The *anti*-BPDE was found to be positive in yielding NIH3T3 transformants derived from the epoxide-modified *ras* plasmid transfection and from TPA-treated cells exposed to the epoxide.

Conclusion

The major organ site for the metabolism and detoxification of AN in humans is the liver. Correlations of the rank-ordered estimates of AN exposure with the liver function and hepatotoxicity tests of the AN-exposed chemical workers have not shown any dose–response effect related to AN. No clinical or epidemiological indication has been identified as the causative agent for human liver cancer from the long-term medical surveillance of the workers in a Louisville chemical plant that AN is a carcinogen like VC. Clinical and epidemiological data of this cohort are consistent with AN being neither a serious hepatotoxin nor a human carcinogen. To provide further support to this conclusion and a better understanding of AN carcinogenesis, we also used a molecular model of Ha-*ras* oncogene activation to evaluate the carcinogenicity of ANO. The exogenous Ha-*ras* gene, in supercoiled plasmid or chromatin (active or quiescent), after treatment with a synthetic epoxide metabolite, was found to be a useful indicator for carcinogenicity. At the cellular level, the epoxide-modified *ras* DNA is expressed in NIH3T3 cells to detect transforming activity. At the molecular level, the promutagenic lesions in plasmid and chromatin Ha-*ras* DNA are mapped by endonuclease S_1. In both comparisons, ANO and *anti*-BPDE appear to be diametrically opposite. This contrast is important to the understanding of carcinogenesis. Recently, Fang et al. (30) developed an in vitro carcinogenesis model in which primary rat hepatocytes were transformed by sequential exposure to SV40 DNA and an activated *ras* oncogene like the EJ-*ras*. Thus, the apparently inadequate modification of Ha-*ras* by ANO and hence inactivity of the DNA may be construed as an indicator of low carcinogenicity or noncarcinogenicity. However, the model study is preliminary, and more refined resolution in sequence analysis of the epoxide-modified Ha-*ras* DNA as well as dose–response effect of ANO will be necessary to confirm this finding. Ultimately, it will be necessary to confirm that human exposure at 1–2 ppm daily (upper limits TWA) and 1–100 ppm in the workplace has no organic basis for carcinogenesis. So far, this oncogene-based, molecular model in combination with the worker surveillance program has provided new information and the possibility of "negative proof" about AN carcinogenicity for the assessment of health risk. This approach may prove fruitful in predicting the relative health risk of environmental carcinogens.

Acknowledgment

This work was supported in part by a grant from National Institute of Environmental Health Sciences (NIEHS) (ES05353).

References

1. *A Recommended Standard for Occupational Exposure to Acrylonitrile;* National Institute for Occupational Safety and Health: Cincinnati, OH, 1978; DHEW (NIOSH) Pub. 78–116.
2. *Environmental Health Criteria, 28, Acrylonitrile;* World Health Organization: Geneva, Switzerland, 1983.
3. McMichael, A. J.; Jelkovic, D. A.; Tyroler, H. A. *Am. N.Y. Acad. Sci.* **1976**, *271*, 125.
4. Waxweiler, R. J.; Smith, A. H.; Tyroler, H. A. *Environ. Health Perspect.* **1981**, *41*, 159.
5. Deizell, A. U.; Monson, R. R. *J. Occup. Med.* **1982**, *24*, 767.

6. Chen, J. L.; Walrath, J.; O'Berg, M. T.; Burke, C. A.; Pell, S. *Am. J. Ind. Med.* **1987**, *11*, 157.
7. Tamburro, C. H.; Creech, J. L., Jr. *J. Univ. Environ. Health*, **1983**, (Suppl.), 37.
8. Joseph, J. T.; Elmore, J. D.; Wong, J. L. *J. Org. Chem.* **1990**, *55*, 471.
9. Roberts A. E.; Lacy, S. A.; Pilon, D.; Turner, M. J., Jr.; Rickert, D. E. *Drug Metab. Dispos.* **1989**, *17*, 481.
10. Recio, L.; Skopek, T. R. *Mutat. Res.* **1988**, *206*, 297.
11. Hogy, L. L; Guengerich, F. P. *Cancer Res.* **1986**, *46*, 3932.
12. Barbacid, M. *Eur. J. Clin. Invest.* **1990**, *20*, 225.
13. Marion, M. J.; Froment, O.; Trdpo, C. *Mol. Carcinog.* **1991**, *4*, 450.
14. Guengerich, F. P. *Chem. Res. Toxicol.* **1992**, *5*, 2.
15. Topal M. D. *Carcinogenesis* **1988**, *9*, 691.
16. Kohwi-Shigematsu, T.; Nelson, J. A. *Mol. Carcinog.* **1988**, *1*, 20.
17. Smith, A. H.; Waxweiler, R. J.; Tyroler, H. A. *Am. J. Epidemiol.* **1980**, *112*, 787.
18. *NIOSH Analytical Methods for Set K*; National Institute of Occupational Safety and Health: Washington, DC, 1976; NTIE Report PB-254-227.
19. Gaganon, Y. T.; Poesner, J. C. *Am. Ind. Hyg. Assoc. J.* **1979**, *40*, 923.
20. Tamburro, C. H.; Greenberg, R. A. *Environ. Health Perspect.* **1981**, *41*, 117.
21. Yuan, B.; Wong, J. L. *Carcinogenesis* **1991**, *12*, 787.
22. Tamburro, C. H. *Semin. Liver Dis.* **1984**, *4*, 159.
23. Spandidos, D. A. *ISI Atlas Sci.: Immunol.* **1988**, 1.
24. Harris, C. C. *Cancer Res. 51*, (Suppl.) **1991**, 5023s.
25. Jeffrey A. M. In *Polycyclic Hydrocarbons and Carcinogenesis*; Harvey, R. G., Ed.; ACS Symposium Series 283; American Chemical Society: Washington, DC, 1985; pp 187-208.
26. Kohwi-Shigematsu, T.; Scribner, N.; Kohwi, Y. *Carcinogenesis* **1988**, *9*, 457.
27. Vousden, K. H.; Bos, J. L.; Marshall, C. J.; Phillips, D. L. *Proc. Natl. Acad. Sci. U.S.A.*, **1986**, *83*, 1222.
28. Ishii, S.; Merlino, G. T.; Pastan, I. *Science (Washington, D.C.)* **1985**, *230*, 1378.
29. Spandidos, D. A.; Nichols, R. A. B.; Wilkie, N. M.; Pintzas, A. *FEBS Lett.* **1988**, *240*, 191.
30. Fang, X. J.; Flowers, M.; Keating, A.; Cameron, R.; Sherman, M. *Cancer Res.* **1992**, *52*, 173.

RECEIVED for review September 3, 1992. ACCEPTED revised manuscript on April 8, 1993.

13

Estimation of Risk of Kidney Dysfunction from Exposure to Cadmium Using Studies of Occupationally Exposed Workers

Elizabeth A. Grossman and Caroline S. Freeman

Occupational Safety and Health Administration, Room N–3507, 200 Constitution Avenue, N.W., Washington, DC 20210

Cadmium can enter the environment through industrial emissions, household waste combustion, burning of fossil fuel, and agricultural fertilizers. The kidney is one of the organs most sensitive to low-level cadmium exposure. Cadmium stored in the kidney can induce dysfunction. Passage of low-molecular-weight proteins in the urine is one of the first signs of dysfunction, but more severe forms can occur. One of the most serious sequelae of cadmium-induced dysfunction, itai-itai or "ouch-ouch" disease, occurred in Japan in an environmentally exposed population. Because studies of occupationally exposed workers may provide a unique estimate of exposure for each member of the cohort, these studies can be used to quantify the risk of kidney dysfunction from exposure to cadmium, and the results may be applied to environmentally exposed populations.

CADMIUM IS A NATURALLY OCCURRING ELEMENT like iron or zinc, but unlike these metals, cadmium has no nutritive value. In this respect, cadmium more closely resembles lead or mercury, but unlike lead or mercury, cadmium is a relatively "modern" metal. Lead and mercury have been used for thousands of years. In contrast, cadmium was not identified as a distinct element until 1817 (*1*).

Cadmium is a soft, silvery, and ductile metal that resembles zinc in most of its chemical properties. It is far less common than lead and somewhat less common than mercury; 0.2 ppm of the earth's crust is composed of cadmium (*1*). Major worldwide uses of cadmium are in electroplating (35%), pigments (25%), plastic stabilizers (15%), and batteries (15%) (*2*).

In nature, cadmium occurs with zinc or lead. It is obtained as a byproduct in the refining of zinc and, to a lesser extent, in the refining of copper and lead from sulfide ores. Ever since metal ores containing cadmium have been refined (i.e., the past several thousand years), cadmium has been a pollutant in the environment. Only in the past 70 years, however, has the amount of cadmium emitted into the environment greatly increased. This is evidenced by findings such as that in Greenland, where snow deposited at one particular site between 1807 and 1917 had cadmium concentrations averaging 0.034 µg/kg of water. Snow deposited at the same site between 1966 and 1971 had cadmium concentrations averaging 0.63 µg/kg, an increase of 1900% (2).

Cadmium enters the environment in a number of ways. Metal smelters and other industries that use cadmium in production emit the metal in the air and discharge it in wastewater. Additional sources of cadmium in the environment are combustion of household waste, which frequently contains products such as plastics and metal scraps; burning of fossil fuels; and use of agricultural fertilizers.

Some cadmium is airborne, but most cadmium is deposited either in water or in soil. Cadmium deposited in water can be absorbed by seafood such as oysters and mussels, thereby entering the food chain. Cadmium deposited in soil can enter the food chain either through plants and crops consumed by humans or through plants and crops consumed by animals that are, in turn, consumed by humans. Cadmium-contaminated water can compound the problem when such water floods crop lands or is used for irrigation.

Inhaled cadmium is more readily absorbed by the body than is ingested cadmium. Approximately 20–25% of cadmium deposited in the lung is systemically absorbed, whereas only about 5% of ingested cadmium is absorbed by the body (3). The majority of absorbed cadmium is deposited in the liver, kidneys, and muscles. After chronic low-level cadmium exposure, about one-sixth of the body burden is found in the liver, one-fifth of the body burden is found in the muscles, and one-half to one-third of the body burden is found in the kidneys—the highest concentrations being found in the renal cortex (3). As the concentration of cadmium to which one is exposed increases, the proportion of cadmium found in the liver increases. Once deposited in these organs, cadmium is retained for a long time. The half-life of cadmium is 5–15 years in the liver, 10 to 30 years in the kidney, and over 30 years in the muscles (3).

The kidney is one of the organs most sensitive to chronic low-level exposure to cadmium. Cadmium is stored in the kidney cortex. When the amount of cadmium stored in the cortex exceeds a threshold level, its presence interferes with the tubular reabsorption of low-molecular-weight proteins, such as β_2 microglobulin and retinol-binding protein. This threshold level is generally considered to be 200 grams of cadmium per gram of kidney cortex, wet weight (4). The condition of passing abnormally high levels of these proteins in urine, known as tubular proteinuria, is an indication that kidney function has been impaired due to damage to the cells lining the proximal tubules. In cadmium-induced renal disease, tubular proteinuria is considered to be one of the earliest signs of renal dysfunction.

After prolonged exposure to cadmium, glomerular proteinuria, a more severe state of kidney dysfunction, may develop in conjunction with tubular proteinuria. This condition is indicated by the presence of high-molecular-weight proteins such as albumin, immunoglobulin G, and a variety of glycoproteins in the urine. It is theorized that cadmium-induced lesions in the tubules and glomeruli result in increased permeability of the glomerulus, which, in turn, results in the loss of high-molecular-weight proteins in the urine.

Cadmium-induced kidney damage is irreversible, and because cadmium is stored in the body and transported from other organs to the kidneys over long periods of time, cessation of exposure may not halt progressive deterioration of kidney function (5). The gravity of cadmium-induced renal dysfunction is compounded by the fact that at present no medical treatment to prevent or reduce the accumulation of cadmium in the kidney exists. As tubular damage progresses, additional signs of kidney damage will emerge. These include glycosuria, characterized by excess glucose in urine; aminoaciduria, characterized by excess amino acid in the urine; phosphaturia, characterized by excess phosphates in the urine; and hypercalciuria, characterized by excess calcium in the urine. Eventually, glomerular function may be affected, as evidenced by the reduction in the glomerular filtration rate.

Studies of Populations with Environmental Exposure to Cadmium

Most people are exposed to relatively low levels of cadmium in the environment. For example, even in highly industrialized areas, airborne cadmium is usually well below 0.1 $\mu g/m^3$ (2). Likewise, estimates of the daily intake of cadmium from food in the United States range from 10 to 51 μg/day (2). There have been cases, however, in which cadmium has contaminated the environment and has caused serious health effects in the exposed population. The majority of these cases have occurred in Japan, where cadmium contaminated the food supply. The most notorious case of cadmium contamination in Japan occurred along the Jinzu River basin in Toyama Prefecture.

The Jinzu River basin is an agricultural region in the central part of Toyama Prefecture. Inhabitants along the river consumed fish from the river, used river water to irrigate the rice paddies, and, until the 1930s, used the river as a source of drinking water. In addition, the river frequently flooded the surrounding farmland.

Upstream from the river basin is the Kamioka Mine, an ore mine in operation for some 370 years. In addition to ore, the mine produced zinc, lead, and cadmium. Wastewater from mine-sludge piles, pit drainage, ore-selection sites, and refineries flowed into the Jinzu River.

During World War II, farmers realized that pollutants in the river water from the mine were damaging their crops, but it was not until the early 1960s that cadmium was identified as the source of damage to the health of the inhabitants along the river. Itai-itai was first recognized as a disease in 1955. Most of the victims were postmenopausal women who had delivered, on average, six children (7). Itai-itai disease, which translates to "ouch-ouch" disease, was characterized by lower back pains and leg muscle pains. Pressure on the bones, especially the femurs, backbone, and ribs, also produced pain. Another characteristic of the disease was a ducklike gait. These conditions would continue for several years until a victim experienced a mild trauma and was then unable to walk. Once the damage was so extensive that a victim was confined to bed, the clinical conditions progressed rapidly. Bones in the extremities and ribs as well as other bones would be susceptible to multiple fractures after very slight trauma, such as coughing. Skeletal deformation would take place, including a marked decrease in body height (7).

At first, itai-itai disease was thought to be a type of osteomalacia attributable to a lack of vitamin D in the diet. However, when rice samples from the region were found to have cadmium concentrations more than 10 times greater than rice samples from other regions of the country, it was postulated that cadmium played an etiological role in the development of the disease. It is now believed that cadmium exposure in combination with malnutrition, specifically a diet low in calcium and vitamin D, were factors associated with the disease. Vitamin D deficiency, which inhibits the body's ability to utilize calcium and phosphates, is well-known to cause rickets in children and osteoporosis and osteomalacia in adults. The production of the biologically active metabolite of vitamin D occurs in the kidney, but when cadmium-induced tubular dysfunction occurs, production of the vitamin D metabolite is reduced (6).

Itai-itai disease was not the only consequence of cadmium pollution in the Jinzu River Basin. An epidemiological study conducted in 1967 found that the prevalence of proteinuria in districts where itai-itai disease was endemic was 50% in men of age 60–69 years and over 60% in women of age 60–69 years. For those living in the endemic districts over 70 years of age, the prevalence of proteinuria was 70–80%. In comparison, the prevalence of proteinuria in nonendemic districts and in borderline districts was approximately 30% or less in men over 60 years of age and less than 40% in women 60–69 years of age. Although women over 70 years of age in the borderline districts had a 60% prevalence of proteinuria, women in this age group who lived in nonendemic districts had a proteinuria prevalence of only approximately 40% (7).

Epidemiological studies in the Jinzu River Basin were followed by studies of residents of other areas of Japan who were potentially exposed to cadmium. The aim of these studies was to identify other potential cadmium polluters, to determine the extent of the possible effects of cadmium, and to search for additional itai-itai disease victims. Although mining was the source of much of the cadmium pollution, entering the food supply through contaminated water, refineries were also an important source of the contamination. For example, in the town of Annaka in Gumma Prefecture, a large zinc refinery contaminated the surrounding farmland, primarily through airborne cadmium. In this area, increased prevalence of tubular proteinuria was observed (8).

Most of the epidemiological studies sought to determine the relationship between cadmium in food and kidney dysfunction. The Japanese Association of Public Health devised a number of methods for estimating cadmium intake from foodstuffs, but most of these methods relied on study subjects' recall of what they ate or on assumptions of average dietary intake. Although these methods are useful for assessing average cadmium intake among groups of people in a population, for example, men, women, and children, they do not allow for differences in cadmium intake among individuals within a group. Thus, although these data can be used to determine whether an association between cadmium exposure and kidney dysfunction exists, the nature of the association, that is, the relationship between dose (i.e., cadmium exposure) and response (i.e., tubular proteinuria) is more difficult to assess.

Studies of Occupationally Exposed Populations

Occupational health studies are particularly useful for determining the etiology of illnesses caused by exposure to toxic substances and for assessing the risk associated with that exposure. One reason for this is that cumulative dose (i.e., the total amount of toxic substance to which a person is exposed) is usually easier to estimate for an occupationally exposed person than it is for an environmentally exposed person. Since the 1960s, the technology has been available to measure most toxic substances, even at very low levels, and an increasing awareness of occupational safety and health has led to improved record-keeping. Thus, in many instances, it is possible to estimate a unique cumulative exposure for each worker in a plant or factory.

Another reason that occupational health studies are particularly useful for assessing risk is that, in general, workers tend to be healthier than the general population. This means that in most working populations, the background incidence of disease tends to be lower than in the general population. Thus, when the incidence of disease is elevated above background in a working population, it is less likely to be attributable to factors other than exposure to the toxic substance.

The Occupational Safety and Health Administration (OSHA) published its *Occupational Exposure to Cadmium; Proposed Rule* in February of 1990 (9). As part of the preamble to that rule, OSHA performed a preliminary quantitative risk assessment. To estimate the risk of kidney dysfunction from exposure to airborne cadmium, OSHA used data from two occupational health studies designed to examine the relationship between cadmium exposure and kidney dysfunction. These studies were chosen for a number of reasons, the most important being that the study authors were able to estimate a unique cumulative cadmium dose for each member of their cohort.

The Ellis Cohort. The first of the studies used by OSHA was by Ellis et al. (10), who studied kidney dysfunction among male workers at a cadmium smelter. The study population consisted of 51 current workers and 31 retired workers with experience in production, nonproduction, office, and laboratory work. Estimates of cumulative cadmium exposure were made for each member of the study population by using historical industrial hygiene data. The chronological record of each worker's job assignments was obtained from personnel files at the smelter. For each worker, cadmium exposure was calculated by multiplying the length of time the worker spent in a given area by the amount of cadmium estimated to have been inhaled in that area during that year (i.e., estimates of exposure were adjusted for respirator usage) and summing over all years in which the worker was exposed, thereby obtaining an estimate of the time-weighted exposure or dose for that worker.

Each cohort member completed a health history questionnaire, took a physical examination, gave specimens for blood and urine tests, and provided 24-h urine samples. In addition, in vivo neutron activation was used to measure the amount of cadmium in the liver and kidney of each worker. The 24-h urine samples were used to determine whether a worker had abnormal kidney function. Kidney function was judged to be abnormal if urinary levels of the low-molecular-weight protein β_2-microglobulin exceeded 200 μg/g creatinine or if total urinary protein levels exceeded 250 mg/g creatinine. Eighteen active workers (35%) and 23 retired workers (74%) were classified as having abnormal kidney function.

Table I presents descriptive statistics for the entire Ellis cohort. In that table, it can be seen that for both active and retired workers, neither age nor duration of exposure is significantly different for workers who had normal kidney function, compared with those who had abnormal kidney function. Only time-weighted cadmium exposure is different for those two groups. For active workers, those with abnormal kidney function had an average time-weighted exposure more than 16 times greater than the aver-

Table I. Descriptive Statistics for a Cohort of 82 Active and Retired Cadmium Smelter Employees

Parameters	Active Workers			
	Normal Kidney Function		Abnormal Kidney Function	
	Mean	(SD)[a]	Mean	(SD)[a]
Number	33		18	
Age (years)	42.6	(13.3)	53.6	(6.8)
Duration of exposure (months)	141	(118)	264	(105)
Time-weighted exposure ($\mu g/m^3$ per year)	105	(9.0)	1690	(2.7)
Renal cadmium ($\mu g/g$)	125	(2.8)	230	(2.0)
Liver cadmium (ppm)	11.3	(2.8)	63.9	(1.5)

Parameters	Retired Workers			
	Normal Kidney Function		Abnormal Kidney Function	
	Mean	(SD)[a]	Mean	(SD)[a]
Number	8		23	
Age (years)	69.0	(8.3)	67.9	(6.9)
Duration of exposure (months)	342	(75)	329	(103)
Time-weighted exposure ($\mu g/m^3$ per year)	379	(3.3)	3143	(3.6)
Renal cadmium ($\mu g/g$)	148	(2.1)	169	(1.7)
Liver cadmium (ppm)	14.0	(3.1)	33.6	(2.9)

[a] Means and standard deviations (SDs) for age and duration of exposure are arithmetic means and SDs. All others are geometric means and SDs.
SOURCE: Reproduced with permission from reference 10. Copyright 1985 Taylor and Francis.

age time-weighted exposure of those with normal kidney function. For retired workers, the average time-weighted exposure of those with abnormal kidney function was more than eight times the average time-weighted exposure of those with normal kidney function.

The Falck Cohort. The second study used by OSHA was by Falck et al. (11), who studied 33 workers at a plant that produces refrigeration compressors with silver-brazed copper fittings. The brazing contained 18–24% cadmium. Compressors were brazed either manually or by an automated process. Estimates of cumulative exposure were made for each worker by using data from air monitoring done by the Michigan Department of Industrial Health. Work history records were obtained for each employee in the study, and a time-weighted exposure for each worker was calculated by multiplying the length of time on each brazing line by the mean estimated exposure for that brazing line.

Each of the 33 workers provided medical histories and spot blood and urine samples. Three workers were dropped from further analysis because of health conditions that affect kidney function. Of the remaining 30 workers, 8 were asked to provide 24-h urine samples, because their urinary glucose, protein, and/or β_2-microglobulin levels exceeded the normal limits (i.e., the 95% tolerance limits) constructed for these variables from the spot urine samples of 41 unexposed workers who served as controls. Glucose, protein, β_2-microglobulin, and creatinine levels were measured in the 24-h urine samples of the eight workers and in 24-h urine samples of seven age-matched male controls. Seven of the eight workers were found to have urinary protein levels in excess of the normal limits (i.e., 95% tolerance limits) constructed for urinary protein from the 24-h samples of the controls; these workers, representing 23% of the entire cohort, were judged to have abnormal kidney function.

Table II presents descriptive statistics for the cohort. The table shows that of age, smoking, and time-weighted exposure, only time-weighted exposure is significantly different between those in the cohort with normal kidney function and those with abnormal kidney function. The average time-weighted exposure for workers with abnormal kidney function is more than two times the average time-weighted exposure for workers with normal kidney function.

Quantification of Risk

OSHA used logistic regression to quantify the risk of tubular proteinuria for any given cumulative cadmium dose. Logistic regression allows one to model the relationship between a continuous variable, in this case cumulative cadmium dose, and a variable with only two possible outcomes, in this case the presence or absence of kidney dysfunction. The model is based on the assumption that the probability of an event (p), in this case the presence of kidney dysfunction, is distributed as a binomial random variable and that the logit function is linear, or

$$\log(p/1-p) = \alpha + \beta x$$

For this specific case, the equation may be written as

$$\text{prob} \log\left[\frac{\text{dysfunction}}{\text{no dysfunction}}\right] = \alpha + \beta \log(\text{dose})$$

Table II. Descriptive Statistics for a Cohort of 30 Employees at a Refrigeration Compressor Production Plant

Parameters	Normal Kidney Function Mean[a] (95% CI)	Abnormal Kidney Function Mean[a] (95% CI)	p-value[b]
Number	23	7	
Age (years)	49	53	.13
	(47, 51)	(51, 55)	
Time-weighted exposure	459	1137	.02
($\mu g/m_3$ per year)	(332, 634)	(741, 1737)	
Smoking habits (pack-years)	14	24	.07
	(9, 19)	(14, 34)	
Urine ratios			
Protein/creatinine (mg/g)	34	246	<.001
	(26, 43)	(132, 456)	
β_2 M/creatinine ($\mu g/g$)[c]	53	6375	<.001
	(31, 90)	(1115; 36,463)	
Cadmium/creatinine	11	16	.07
($\mu g/g$)	(10, 13)	(8, 36)	
Serum ratios			
Creatinine/serum	1.1	1.4	.003
(mg/100 mL)	(1, 1.2)	(1.2, 1.7)	
β_2 M/serum ($\mu g/mL$)	2	2.3	.32
	(1.6, 2.4)	(1.8, 2.8)	

[a] Means for age and smoking habits are arithmetic means; all others are geometric means. Ninety-five percent confidence intervals (95% CIs) are constructed from arithmetic standard deviations for age and smoking; all others are constructed from geometric standard deviations.
[b] p-value is associated with a test of difference between group means.
[c] β_2 M is β_2 microglobulin.
SOURCE: Reproduced with permission from reference 11. Copyright 1983 Wiley-Liss.

The logistic regression model may be used not only to model the relationship between dose and dysfunction but also to estimate the risk of dysfunction (i.e., the probability of response) for any given dose. Once the model has been fit to the data and the parameters α and β have been estimated, the risk of kidney dysfunction for any given dose may be calculated by using the equation

$$\text{prob dysfunction} = \frac{\text{dose}^b}{\exp(a) + \text{dose}^b}$$

where a and b are the estimated values of α and β, respectively.

The risks associated with a variety of cumulative airborne exposures (i.e., doses) are presented in Table III. The 8-h time-weighted average (TWA) dose represents the level of airborne cadmium to which workers are exposed during an 8-h shift. The cumulative dose, expressed in $\mu g/m^3$ per year, represents the total dose a worker receives. The cumulative dose of a worker exposed to an 8-h cadmium TWA of 5 $\mu g/m^3$ for 45 years would be 225 $\mu g/m^3$ − years (5 × 45 = 225). For a worker exposed to an 8-h cadmium TWA of 10 $\mu g/m^3$ for 20 years, the cumulative dose would be 200 $\mu g/m^3$ per year (10 × 20 = 200).

One member of the Ellis cohort, an 80-year-old retired office worker, was classified as having kidney dysfunction, although his cumulative cadmium exposure was low. His level of urinary β_2-microglobulin was just slightly elevated over the 200 $\mu g/g$ creatinine limit, and he was the only member of the cohort with abnormal kidney function at a cumulative exposure level of less than 400 $\mu g/m^3$ per year. Because one observation may be very influential in a logistic regression, OSHA refit the logistic regression model to the Ellis data, excluding this one case. The risks estimated from this modified Ellis model are also presented in Table III.

OSHA fit a logistic regression model to the data from Falck et al. (11), and the risks estimated from this model are presented with the risk estimates from the Ellis model in Table III. The model used by OSHA is very simple and does not take into account the role of other factors that may be confounding the relationship between cadmium exposure and kidney dysfunction. Two such variables are age and smoking. Urinary protein levels increase with age, and cigarettes, which contain approximately 2 μg of cadmium apiece, represent an additional source of cadmium exposure. Falck provided OSHA with data on these two potential confounders for each member of his

Table III. Estimates of Kidney Dysfunction per 10,000 Workers with 45 Years of Occupational Exposure to Cadmium

8 Hour TWA Dose ($\mu g/m^3$)	Cumulative Dose ($\mu g/m^3$ per year)	Incidence of Kidney Dysfunction		
		Ellis Model[a]	Modified Ellis Model[b]	Falck Model[c]
1	45	261	83	1
5	225	1646	981	90
10	450	3177	2467	589
20	900	5237	4965	3005
40	1800	7220	7480	7467
50	2250	7740	8089	8457
100	4500	8900	9272	9741

[a] The estimated parameters for this model are $a = -8.34$ and $b = 1.24$.
[b] The estimated parameters for this model are $a = -19.75$ and $b = 2.78$.
[c] The estimated parameters for this model are $a = -10.83$ and $b = 1.59$.

cohort. In its analysis of the data, OSHA found that only the cumulative cadmium dose made a significant contribution to the predictive value of the model. In other words, neither age nor smoking could account for the relationship between cadmium dose and kidney dysfunction observed in the Falck cohort.

Table III shows that the highest estimates of risk are predicted from the model applied to the Ellis data. This finding is reasonable because the Ellis data set included retired workers, and age is a factor in loss of kidney function. When the retired worker with dysfunction but little cadmium exposure is excluded from the analysis, the risks at the lower doses decline. This finding indicates that this one data point was indeed influential in the fit of the model. At the lower doses, the model applied to the Falck data predicts the lowest risks, but by 1800 μg/m^3 per year, the risks predicted from the Falck data are the same as those predicted from the Ellis data. At higher doses, the Falck data predict higher risk than the Ellis data, either with or without the outlier.

Another way to compare the risks predicted from these data sets is to view a graph of the logistic models fit to these data. Figure 1 shows the fitted logistic regression line for the entire Ellis cohort, for the modified Ellis data (i.e., the Ellis data with the outlier removed), and for the Falck data. On the horizontal axis, cumulative dose is expressed in μg/m^3 per year. The vertical axis shows the probability of having kidney dysfunction. The fitted lines for all three models have the same basic shape, but as was seen in Table III, the Falck model gives the lowest risks at the lowest doses and the highest risks at the highest doses.

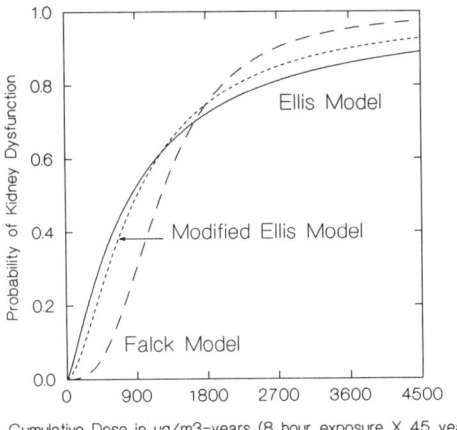

Figure 1. Probability of kidney dysfunction by cumulative cadmium dose estimated from the Ellis data, the modified Ellis data, and the Falck data.

Uncertainty

Although there is uncertainty in any assessment of risk, when the results are corroborated by other epidemiological studies, confidence in one's estimates of risk increases. Five other studies of occupationally exposed workers support OSHA's assessment of risk (12–16).

Figure 2 (17) is another plot of the kidney risk model fitted by OSHA, but it also plots the observed prevalences of kidney dysfunction from seven studies of workers with occupational exposure to airborne cadmium, including the Ellis and Falck studies (10–16). In addition, a metabolic model derived by Kjellstrom and Nordberg (18) is presented in Figure 2. The observed data in this figure show a pattern between dose and response similar to that predicted by OSHA in its risk assessment. The prevalence of kidney dysfunction increases sharply at cumulative exposures above 500 $\mu g/m^3$ per year. The OSHA model generally follows the upper range of the observed data and agrees well with the metabolic model. This agreement increases confidence in the risk estimates derived from the OSHA risk assessment and increases the likelihood that the logistic regression models provide plausible estimates of the risk of kidney dysfunction from exposure to airborne cadmium.

In summary, we can see that occupational health studies can be useful tools for assessing risk from environmental exposure to toxic substances, because these studies provide dose data for each member of the cohort, thereby allowing risk to be estimated for an infinite number of dose levels.

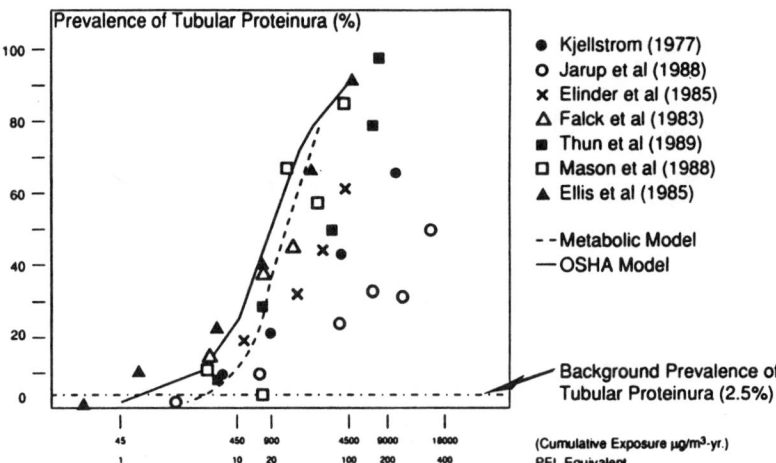

Figure 2. Prevalence of tubular proteinuria by cumulative exposure to cadmium in seven cross-sectional studies compared with prediction by the OSHA risk assessment and by the Kjellstrom metabolic model. (Reproduced with permission from reference 17. Copyright 1991 Wiley-Liss.)

Acknowledgment

The authors thank Ed Stern for his technical assistance in the preparation of this paper.

References

1. Hammond, P. B.; Beliles, R. P. In *Casarett and Doull's Toxicology; The Basic Science of Poisons*, 2nd ed.; Doull, J.; Klaassen, C. D.; Amdur, M. O., Eds.; Macmillan: New York, 1980.
2. Elinder C. G. In *Cadmium and Health: A Toxicological and Epidemiological Appraisal;* Friberg, L.; Elinder, C. G.; Kjellstrom, T.; Nordberg, G. F., Eds.; CRC: Boca Raton, FL, 1985; Vol. 1.
3. Nordberg, G. F.; Kjellstrom, T.; Nordberg, M. In *Cadmium and Health: A Toxicological and Epidemiological Appraisal;* Friberg, L.; Elinder, C. G.; Kjellstrom, T.; Nordberg, G. F., Eds.; CRC: Boca Raton, FL, 1985, Vol. 1.
4. Lauwerys, R. R.; Malcom, D. *Health Maintenance of Workers Exposed to Cadmium: A Guide for Physicians*; Cadmium Council: New York, 1985.
5. Piscator, M. *Environ. Health Perspect.* **1984**, *54*, 175–179.
6. Kjellstrom, T. In *Cadmium and Health: A Toxicological and Epidemiological Appraisal;* Friberg, L.; Elinder, C. G.; Kjellstrom, T.; Nordberg, G. F., Eds.; CRC: Boca Raton, FL, 1986; Vol. 2.
7. *Cadmium in the Environment*, 2nd ed.; Friberg, L.; Elinder, C. G.; Kjellstrom, T.; Nordberg, G. F., Eds.; CRC: Boca Raton, FL, 1974.
8. Fukushima, I. In *Cadmium Studies in Japan: A Review;* Tsuchiya, D., Ed.; North-Holland Biomedical: Amsterdam, The Netherlands, 1978.
9. *Occupational Exposure to Cadmium; Proposed Rule;* Occupational Safety and Health Administration: Washington, DC, February 6, 1990; 55FR40520.
10. Ellis, K. J.; Cohn, S. H.; Smith, T. J. *J. Toxicol. Environ. Health* **1985**, *15*, 173–187.
11. Falck, F. Y.; Fine, L. J.; Smith, R. G.; McClatchey, K. D.; Annesley, T.; England, B.; Schork, A. M. *Am. J. Ind. Med.* **1983**, *4*, 541–549.
12. Elinder, C. G.; Edling, C.; Lindberg, E.; Kagedal, B.; Vesterberg, O. *Br. J. Ind. Med.* **1985**, *42*, 754–760.
13. Jarup, J.; Elinder C. G.; Spang, G. *Int. Arch. Occup. Environ. Health* **1988**, *60*, 223–229.
14. Kjellstrom, T.; Borin, P.; Rahnster, B. *Environ. Res.* **1977**, *13*, 303–317.
15. Mason, H. J.; Davison, A. G.; Wright, A. L.; Guthrie, C. J. G.; Fayers, P. M.; Venables, K. M.; Smith, N. J.; Chettle, D. R.; Franklin, D. M.; Scott, M. C.; Holden, H.; Gompertz, D.; Newman-Taylor, A. J. *Br. J. Ind. Med.* **1988**, *45*, 793–802.
16. Thun, M. J.; Osorio, A. M.; Schober, S.; Hannon, W. H.; Lewis, B.; Halperin, W. *Br. J. Ind. Med.* **1989**, *46*, 689–697.
17. Thun, M. J.; Elinder, C. G.; Friberg, L. *Am. J. Ind. Med.* **1991**, *20*, 629–642.
18. Kjellstrom, T.; Nordberg, G. F. In *Cadmium and Health: A Toxicological and Epidemiological Appraisal;* Friberg, L.; Elinder, C. G.; Kjellstrom, T.; Nordberg, G. F., Eds.; CRC: Boca Raton, FL, 1985; Vol. 1.

RECEIVED for review September 28, 1992. ACCEPTED revised manuscript January 26, 1993.

14

Estimating Malathion Doses in California's Medfly Eradication Campaign Using a Physiologically Based Pharmacokinetic Model

Michael H. Dong[1], William M. Draper[2], Paul J. Papanek, Jr.[3], John H. Ross[1], Kimberley A. Woloshin[3], and Robert D. Stephens[2]

[1]Worker Health and Safety Branch, California Department of Pesticide Regulation, 1020 N Street, Sacramento, CA 95814
[2]Hazardous Materials Laboratory, California Department of Health Services, 2151 Berkeley Way, Berkeley, CA 94704
[3]Toxics Epidemiology Program, Los Angeles County Department of Health Services, 2615 South Grand Avenue, Los Angeles, CA 90007

> *Physiologically based pharmacokinetic (PB-PK) models are designed to simulate the body as a series of tissue compartments, across which a chemical is absorbed, distributed, metabolized, and excreted in accord with pharmacokinetic rate laws. Well-constructed PB-PK models can be powerful tools for interpretation of biomarker data. This chapter demonstrates such use in estimating the absorbed malathion doses in subjects potentially exposed during an urban pesticide application. Single urine samples were collected from subjects within 48 h of a potential exposure for determination of malathion dicarboxylic acid, one of the pesticide's major metabolites. Subjects also responded to a questionnaire that provided a brief description of the circumstances and timing of their exposure. This PB-PK simulation suggests that for the adults (70 kg) and the children (14–34 kg), the highest absorbed doses were 1.3 and 0.4 mg, respectively.*

CHARACTERIZING CHEMICAL EXPOSURES IN A STUDY POPULATION is one of the most difficult challenges in environmental epidemiology, a fact that has been emphasized in a number of chapters in this volume. The utility of biomarkers as probes of chemical exposure has been widely appreciated and, as a result, the technology for measuring markers in biological samples has advanced significantly. In particular, analytical methods of determining xenobiotic metabolites in biological fluids (i.e., biomarkers of internal dose) are increasingly sensitive and reliable. Successful application of the new assays in epidemiological studies is not straightforward, however, because exact knowledge of the concentration of a biomarker in a sample does not translate to a dose (the parameter of greatest interest to the epidemiologist). Physiologically based pharmacokinetic (PB-PK) models are designed to treat the body as a series of tissue compartments, across which a chemical is absorbed, distributed, metabolized, and excreted in accord with pharmacokinetic rate laws. Well-constructed PB-PK models can be powerful tools for interpretation of biomarker data, as demonstrated here in estimating absorbed malathion doses in subjects allegedly exposed to aerial sprays during an urban pesticide application.

Malathion (S-1,2-bis[ethoxycarbonyl]ethyl O,O-dimethyl phosphorodithioate) is an insecticide that has been used widely in the United States to control mosquitoes, flies, and household insects since the early 1960s. This insecticide is a member of the organophosphorous group and, as such, can cause acute cholinergic reactions in humans at high dosages. Malathion is readily metabolized by microsomal enzymes through oxidation of the P=S bond to P=O, followed by hydrolysis of the phosphate ester. However, the predominant metabolic pathway in humans is deesterification of the ethyl succinate esters; two of its major urinary metabolites in humans are the monocarboxylic acid (MCA) and dicarboxylic acid (DCA). Gallo and Lawryk (1) recently provided a thorough review of the toxicology and metabolism of this pesticide in mammals.

As part of the recent Mediterranean fruit fly (Medfly) eradication campaign in California, aerial sprays of malathion mixed with a corn-syrup protein bait were applied repeatedly (up to a dozen times) over portions of southern California during the fall of 1989 and continuing through July 1990. Much of the area sprayed was urbanized and included large population centers. Public concern about the health risks of these aerial applications was high, despite reassurances by state and local health officials. In an effort to address this widespread community concern, state health officials cooperated to collaborate with the Los Angeles County Department of Health Services staff to study doses of malathion in local residents alleging exposure to the aerial sprays. This collaborative effort led to the recruitment of over 60 local residents and agricultural workers for a study in which malathion's acid metabolites in urine were monitored. Malathion metabolites in human urine are detectable at levels as low as a few parts per billion (2–4), thus having the potential to quantify exposure to microgram quantities of the pesticide.

The purpose of this chapter is to illustrate how PB-PK modeling can be used in an epidemiological study to estimate the total absorbed dose of malathion when coupled with internal-dose biomarker data, in this case urinary malathion acid metabolite concentrations. To date, the development and validation of PB-PK models has largely been accomplished by toxicologists studying experimental animals. This simulation study is unique, however, in that it involves local residents and agricultural workers with probable exposures incurred under "real world" conditions.

Materials and Methods

Study Cases and Exposure Scenario. For the Medfly eradication campaign in southern California during 1989–1990, a 20% mixture of technical grade malathion (with 95% purity) diluted in a corn-syrup protein bait was applied over many nights by helicopter over several hundred square miles, at a rate of 2.3 oz active ingredient per acre (equivalent to approximately $2\mu g/cm^2$). The aerial applications were found to be uniform, as the deposition measured on test cards placed throughout spray areas rarely varied by more than 50% of the application rate. Spray droplets ranged in diameter from about 100 μm up to about 1–2 mm. By the following morning, the sprayed droplets typically had dried and hardened and were difficult to dislodge from flat surfaces. Malathion concentrations in outdoor air ranged up to a few micrograms per cubic meter of air.

A total of 67 individuals participated in the Los Angeles biomonitoring study. These included 30 women (29 residents and 1 agricultural worker), 20 men (13 residents and 7 agricultural workers), and 17 children (8 girls and 9 boys). Participants (or parents of the children) recruited were asked during an interview to respond to a questionnaire about their age, gender, circumstances of exposure, duration of exposure, and prevailing symptoms. The general profile of the participants and their exposure experience were previously described in detail by Papanek and Woloshin (5). Participants were instructed to collect a urine specimen at home in any ordinary clean glass jar and to freeze the specimen immediately. They were also asked to label the specimens with the date and time of collection. The specimens were then collected by county health department staff, thawed in the cold, and transferred to uniform glass jars, which were frozen again. These refrozen specimens were later sent to the Hazardous Materials Laboratory (HML) of the California Department of Health Services in Berkeley for analysis. In addition, 24 specimens were split to provide the Pacific Toxicology Laboratories in Los Angeles with duplicates to be analyzed for quality control purposes.

For the case study described here, only the 11 subjects with detectable malathion acid metabolites in their urine were included for PB-PK simulation. These 11 study cases, together with their estimated level of total acid metabolites (i.e., MCA + DCA), are listed in Table I. These 11 subjects all reported either to have been outdoors at night directly under an aerial application of the malathion bait or to have had extensive and direct skin con-

Table I. Excretion of Urinary Malathion Acid Metabolites by Subjects Potentially Exposed to Medfly Eradication Sprays[a]

Subject[b]	Case ID	Age	Creatinine (g/L urine)	Acid Metabolites (μg/L urine)[c]
Adult residents (70 kg)				
A	03	60	1.7	20.7
B	04	48	2.1	81.0
C	39	60	1.9	21.6
D	63	51	1.2	147.0
Agricultural workers (70 kg)				
E	AG49	39	1.2	45.0
F	AG52	45	1.3	81.0
G	AG53	57	1.0	40.0[d]
Younger children (14 kg)				
H	48	2	0.2	225.0
I	61	3	1.4	24.9
Older children (35 kg)				
J	31	10	2.3	75.0
K	60	5	1.2	156.0

[a]Presumably from direct spray or skin contact.
[b]The time lapse from first dermal contact until urine collection was reported to be within 12 h for all study cases, except subjects B, C, and D; according to their questionnaire, the time lapses for subjects B, C, and D were within 24–36 h, 12–36 h, and 36–48 h, respectively.
[c]Based on the dicarboxylic acid metabolites measured by the HML of the California Department of Health Services; total mono- and dicarboxylic acid metabolites were estimated to be three times the amount of dicarboxylic acid metabolites measured.
[d]Based on the total mono- and dicarboxylic acid metabolites measured by the Pacific Toxicology Laboratories in Los Angeles, whose analytical results were used primarily for quality-control purposes.

tact with a sprayed surface, such as grass or backyard foliage. All other subjects with less exposure, such as those merely residing in or walking through a sprayed area, were found to have undetectable levels of urinary metabolites. Also included in Table I are the urinary creatinine levels and subjects' ages. As mentioned in the table footnote, only DCA was measured by HML. The urine samples were collected and analyzed before a commitment was made to estimate also the absorbed doses through PB-PK simulation, which at this time can make use of only the total acid metabolites since the required metabolic rate constants (V_{max}) and Michaelis constant (K_M) for simulation are unavailable for the individual DCA or MCA. The MCA for each of the

11 study cases thus was estimated as two times that of the measured DCA in order to account for the total acid metabolites excreted.

According to Bradway and Shafik (2), over half of the malathion metabolites excreted in human urine appear in the form of MCA or DCA. A malathion clearance study conducted recently by the California Department of Pesticide Regulation (DPR) also indicated that the ratio of MCA to DCA metabolites in human urine changed over time (6). This ratio ranged from approximately 3:1 at 6-h to 1:1 at 12 h following exposure, which suggested sequential deesterification. The MCA-to-DCA ratio used here was considered to be valid, in that the mean MCA (provided by the reference laboratory) among the quality control samples was approximately 1.9 times higher than their mean DCA.

PB-PK Simulation. For the past decade, numerous investigators have used PB-PK models to predict tissue dose in animals and humans (7–25). These models can be highly isomorphic with the physiological and biochemical system of a specific mammalian species and thus may be useful to both toxicologists and epidemiologists. PB-PK models are defined by a set of mathematical equations used to simulate the time course of a chemical's disposition in several preselected tissue compartments. Each compartment has its own characteristic blood flow, volume, tissue-blood partition coefficient, and metabolic- or clearance-rate constants that together explain the chemical's disposition in that region. A comprehensive discussion of the details of constructing PB-PK models has been provided by Bischoff (26), Gibaldi and Perrier (27), and Andersen (28). The two basic steps in the construction of a PB-PK model are (1) choice of body regions and (2) postulation of a set of mathematical equations that will adequately relate the chemical's disposition in the preselected regions.

The basic structure of a PB-PK model for dermal exposure is outlined in Figure 1. The dermal-exposure model in Figure 1 shows that there is a series of mass-balance differential equations needed to account for the time course of the chemical's disposition occurring in the tissue compartments. The model assumes that where the skin is exposed to a chemical, a portion diffuses into the skin, and the rest is either lost to the atmosphere (as by evaporation) or remains on the skin. The portion that has been absorbed into the skin is further assumed to be gradually distributed to the various tissue compartments via circulation. In general, the amount in a particular tissue compartment is affected by many of the following biological variables: (1) the amount of the chemical available in the circulation at the time in question; (2) the cardiac output and the blood flow to the tissue involved; (3) the rates of clearance and metabolism, if they take place; and (4) the partition coefficient between the tissue and the blood involved, which is also defined as the solubility or concentration of the chemical in the tissue, com-

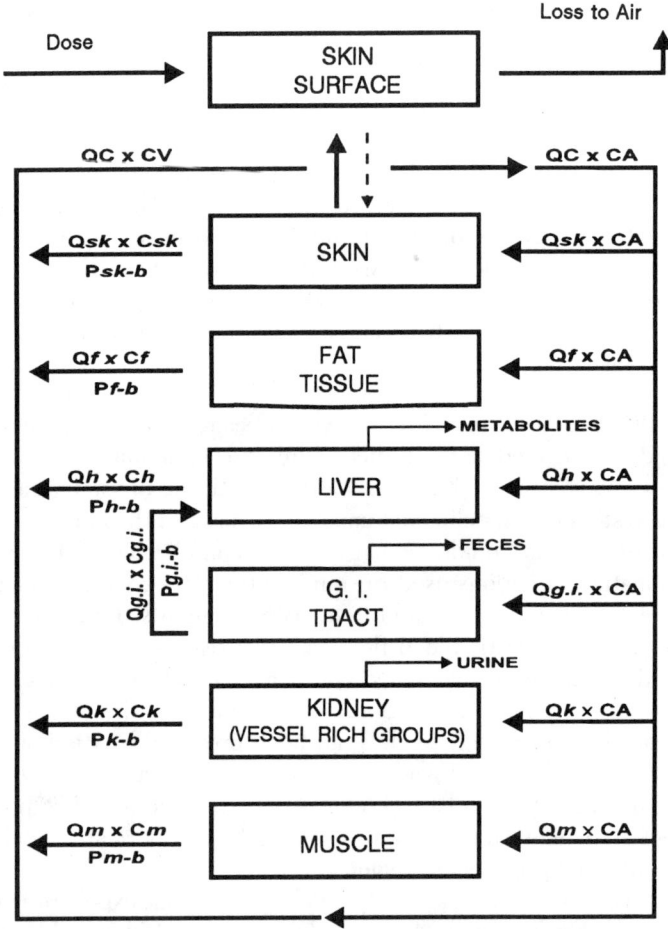

Figure 1. A typical PB-PK model for dermal exposure (see box for explanation of notations used).

Q_i is blood flow to tissue i; P_{i-b} is tissue i − blood partition coefficient; C_i is concentration in tissue i; QC is cardiac output; CV is mixed venous concentration; CA is mixed arterial concentration; and CA is equal to CV.

pared with that in the blood at a steady state. The variables in the last two categories are highly specific to the chemical under investigation.

Many of the differential equations that are typically used in a model for dermal exposure (*see* box) have been used in the PB-PK models for dermal absorption of pesticides by Knaak et al. (20) and of organic vapors by McDougal et al. (9, 21). Equation 6 relies on the widely accepted assumption that venous liver concentrations, rather than liver concentrations, are used to derive the V_{max} and K_M values involved. The kinetic equations listed in the box consist of nonlinear terms and hence cannot be directly integrated. They can be solved indiscriminately, however, with numerical procedures that have been written to approximate an analytical solution.

One numerical procedure that has been considered to offer the most accurate (but also more laborious) integration approximation is the Runge–Kutta method, the computational details of which are readily available in many modern elementary textbooks on differential equations. The algorithm for this numerical procedure can be implemented with any computer programming language such as FORTRAN, BASIC(A), or PASCAL. Within DPR, this algorithm has been written in BASIC(A) for PB-PK simulation on an IBM-compatible microcomputer (29). More sophisticated simulation packages, such as SimuSolv, SCoP, and STELLA, are also available for solving differential equations formulated by the analyst. To date, however, no commercially available simulation programs have been written specifically for PB-PK simulation of dermal exposure.

Table II lists all the physiological and biochemical values that were used in the PB-PK model constructed here specifically for dermal exposure to malathion. Many of these parameter values were also used in two earlier related studies by the California Environmental Protection Agency, Office of Environmental Health Hazard Assessment (OEHHA), where justification for their use was given (30, 31). Further elaboration on the use of these values is provided in the next section. In addition to equation 6 noted earlier, another modification made here to the OEHHA model was the assumption that malathion leaving the gastrointestinal tract compartment would circulate into the liver compartment. This additional modification is consistent with common practice in which the gastrointestinal tract is treated as a separate compartment and is reflected in Figure 1 and in equation 7.

The objective of this PB-PK modeling was to simulate model predictions of the cumulative urinary excretion of malathion acid metabolites to the amounts equivalent to those estimated from the biomonitoring data. Once there was a good approximation (match) between the simulated and the observed values, the total absorbed dose of malathion would be calculated by summing the amounts simulated for all internal tissues, including those remaining in the skin and those excreted. The observed urinary excretions for the 11 study cases were first calculated for a 24-h accumulation by multiplying the estimated urinary level of acid metabolites in Table I by the assumed daily urine output volume. The 24-h accumulation was then adjusted for the two (shortest and longest) time from exposure to urine collection assumed for each study case. The daily urine output volumes for the adults, younger children, and older children were assumed to be 1200, 500, and 800 mL, respectively (32). Because the creatinine levels measured for the study cases indicate that the daily urine output volumes tended to be overestimated

Equations Typically Used in a PB-PK Model for Dermal Exposure[a]

$$CA = \{(Q_{sk} \times C_{sk}/P_{sk\text{-}b}) + (Q_f \times C_f/P_{f\text{-}b}) + (Q_h \times C_h/P_{h\text{-}b}) + (Q_k \times C_k/P_{k\text{-}b}) + (Q_m \times C_m/P_{m\text{-}b})\}/QC \quad (1)$$

Mixed arterial (CA) concentration is the sum of the amounts eliminated by the individual compartments divided by the cardiac output QC (i.e., by the sum of the individual Q_i). The amount eliminated in each compartment i is denoted by $Q_i \times C_i/P_{i\text{-}b}$, where Q_i = blood flow to tissue i, C_i = concentration in tissue i, and $P_{i\text{-}b}$ = tissue i/blood partition coefficient. (Throughout this box, the following notations are used: sk = skin; f = fat; h = hepatic; k = kidney; g.i. = gastrointestinal tract; m = muscle; surf = skin surface; met = metabolite; u = urine; fec = feces; and AMT = amount.)

$$dAMT_{surf}/dt = K_{sp} \times A \times (C_{sk}/P_{sk\text{-}a} - C_{exp}) - K_a \times AMT_{surf} \quad (2)$$

The amount *on* the skin (AMT_{surf}) changes over time as a function of three events: (1) the amount diffused from *inside* to *outside* of the skin (*although at times this amount is negligible*); (2) the amount absorbed *into* the skin; and (3) the amount lost to the air. (K_{sp} = skin permeability constant; A = skin surface area exposed; $P_{sk\text{-}a}$ = skin–air partition coefficient; K_a = evaporation constant; and C_{exp} = concentration of dose applied *topically*.) (Where C_{exp} is averaged *air* concentration, both this and Equation 3 will no longer be applicable, because in that case the applied dose will not diminish over time.)

$$dAMT_{air}/dt = K_a \times AMT_{surf} \quad (3)$$

The amount lost to the air is a function of $K_a \times AMT_{surf}$

$$dAMT_{sk}/dt = K_{sp} \times A \times (C_{exp} - C_{sk}/P_{sk\text{-}a}) + Q_{sk} \times (CA - C_{sk}/P_{sk\text{-}b}) \quad (4)$$

The amount absorbed into the skin is a function of three events: (1) the amount diffused *into* the skin; (2) the amount diffused from *inside* to *outside* of the skin; and (3) the amount of difference between that perfused to and that eliminated in the skin tissue.

$$dAMT_f/dt = Q_f \times (CA - C_f/P_{f-b}) \quad (5)$$

The amount in fat is related directly to the amount of difference between that perfused to and that eliminated in the fat tissue.

$$dAMT_{met}/dt = (V_{max} \times C_h)/\{(K_M \times P_{h\text{-}b}) + C_h\} \quad (6)$$

This metabolism rate is based on the well-known Michaelis–Menten equation; for some chemicals, this equation may occur in a tissue organ other than the hepatic system or may take another form, such as a first-order reaction.

Equations Typically Used in a PB-PK Model for Dermal Exposure[a]—*Continued.*

$$dAMT_h/dt = Q_h \times (CA - C_h/P_{h\text{-}b}) \\ + (Q_{g.i.} \times C_{g.i.}/P_{g.i.\text{-}b}) - dAMT_{met}/dt \quad (7)$$

$$dAMT_u/dt = K_u \times AMT_k \quad (K_u = \text{urinary constant}) \quad (8)$$

$$dAMT_k/dt = Q_k \times (CA - C_k/P_{k\text{-}b}) - K_u \times AMT_k \quad (9)$$

$$dAMT_{g.i.}/dt = Q_{g.i.} \times (CA - C_{g.i.}/P_{g.i.\text{-}b}) \quad (10)$$
$$- K_{fec} \times GI \quad (K_{fec} = \text{fecal constant})$$

$$dAMT_{fec}/dt = K_{fec} \times GI \quad (11)$$

$$dAMT_m/dt = Q_m \times (CA - C_m/P_{m\text{-}b}) \quad (12)$$

[a]These equations are summarized graphically in Figure 1. Equations 7 through 12 are not elaborated here, because on reviewing the first few equations, the reader should find their interpretations all to be repetitive. For dermal exposure, mixed venous concentration (as denoted by CV in Figure 1) is assumed to be approximately equal to CA.

(perhaps with the exception for the agricultural workers), their 24-h accumulations and absorbed doses were likely to be overestimated as well. Several investigators (33, 34) have recently found the excretion rate of urinary creatinine to change over time. It was primarily because of these observations that the creatinine levels were not used here to calculate the daily urine output volumes.

Results and Discussion

The amounts of malathion and of its acid metabolites simulated for the various tissue groups for subject B are presented in Table III. These results exemplify the total absorbed dose of malathion simulated for the longest assumed time lapse from exposure to urine collection for subject B, whose observed malathion acid metabolites in urine accumulated up to 36 h were estimated to be (81 µg/L × 1.2 L/24 h × 36 h =) 145.8 µg (*see* footnotes in Table I and the box for assumptions). As shown in Table III, the accumulation of acid metabolites simulated at 36 h was 145.79 µg. In order for the body to produce and excrete the acid metabolites in this amount in 36 h under the physiological and biochemical constraints specified by the model, the amounts of malathion and its metabolites accumulated in the various tissue compartments should approximate those specified in the table. The total absorbed dose of malathion for this case was estimated to be 0.56 mg, which is simply the sum of the amounts of malathion and of its acid

Table II. Typical Physiological and Biochemical Values Used in PB-BK Human Models for Dermal Exposure to Malathion[a]

Parameters	Adult (70 kg)	Younger Child (14 kg)	Older Child (35 kg)
A. Tissue volumes (L)			
Fat	10.0	2.0	5.0
Intestine	2.4	0.48	1.2
Kidney[b]	2.7	0.54	1.35
Liver	1.5	0.30	0.75
Muscle	30.0	6.0	15.0
Skin	2.6	0.52	1.3
B. Tissue perfusion rates (L/min)			
Fat	0.2	0.06	0.12
Intestine	1.2	0.36	0.71
Kidney[b]	2.25	0.68	1.34
Liver	1.5	0.45	0.89
Muscle	1.2	0.36	0.71
Skin	0.125	0.04	0.07
Cardiac output	6.475	1.95	3.84
C. Hydrolytic liver metabolism[c]			
V_{max} (mole/min)	4.89×10^{-4}	1.46×10^{-4}	2.91×10^{-4}
K_m (mole/L)	1.35×10^{-4}	1.35×10^{-4}	1.35×10^{-4}
D. Other kinetic parameters (min^{-1})			
Skin permeability constant	1.0×10^{-4}	1.0×10^{-4}	1.0×10^{-4}
Evaporation constant	20.0	20.0	20.0
Urinary constant	1.0×10^{-4}	1.0×10^{-4}	1.0×10^{-4}
Fecal constant	0.1	0.1	0.1
E. Tissue/blood partition coefficient			
Fat		775.0	
Intestine		15.0	
Kidney[b]		17.0	
Liver		33.6	
Muscle		22.8	
Skin		25.0	

[a]Based on those adopted by the California Environmental Protection Agency OEHHA (30, 31).
[b]Including other vessel rich groups (brain, heart, and lungs).
[c]Available only for total mono- and dicarboxylic acid metabolites.

metabolites simulated for all internal tissues at any time after 8 h from first dermal contact (after the dermal dose presumably was washed off).

In Table III, the disposition of malathion in the various tissue compartments for subject B was for preselected hourly intervals only. The PB-PK model actually simulated the amount of chemical in each tissue at 0.1-min (6-sec) intervals. The output American Standard Code for Information Interchange (ASCII) file from the computer program, on the other hand, listed these serial amounts at every 10-min interval. Although in Table III the amounts listed for some of the tissues decreased over time as a result of redistribution, all listed amounts actually represented an accumulation or an account of the chemical or its metabolites available up to the specified time interval, rather than that present at the specified time. Figure 2 provides a graphic view of this output for the acid and other metabolites excreted by subject B. As shown in Figure 2, the ratio of acid metabolites to other malathion metabolites for this individual was approximately 1:2. In addition to their (cumulative) total, the excretion rate of malathion acid metabolites may also be determined from the simulation data listed in the output file. The excretion rates of malathion acid metabolites calculated at various time intervals for subject B are depicted in Figure 3. As shown, the excretion rate peaked at approximately 9 h after first dermal contact.

The total absorbed doses of malathion simulated for the 11 study cases are summarized in Table IV. This summary table shows that subject D experienced the highest dose, although this individual did not have the highest level of urinary metabolites. This simulation result was not inconsistent with general expectation, however, in that the subject's daily urine output was assumed to be 1200 mL. Because of age difference, the daily urine outputs for subjects H and K were assumed to be 500 and 800 mL, respectively. Thus, on a cumulative basis, subject D should have higher urinary acid metabolite output (which would lead to the estimation of a higher absorbed dose), even though subjects H and K had the highest metabolite levels detected.

Overall, the result simulated from this PB-PK modeling were found to be consistent with those available in the literature. In this case study, the acid metabolites in the liver were assumed to be excreted primarily into the urine shortly after they were produced. According to Bradway and Shafik (2), MCA and DCA excreted in human urine could account for 57% of the total urinary malathion metabolites. Table III suggests that the acid metabolites accumulated through simulation were approximately 36% of the total malathion metabolites excreted in urine. Although there appeared to be a significant difference between the two studies in the percentage of acid metabolites excreted, the lower percentage of acid metabolites obtained from this PB-PK simulation was not unexpected. The PB-PK model constructed here was for dermal exposure, whereas the observation by Bradway and Shafik involved oral exposure based on a single suicide case. As can be seen

Table III. PB-PK Simulation of Amounts of Malathion in Tissues for Subject B over a One-Week (168-h) Period[a]

Hours Since Contact[b]	On Skin	To Air	In Skin	Fat	Liver	GI Tract	Feces	Kidney	Urine	Muscle	Acid Metabolites[c]
0	13.486	0.000	0.000	0.000	0.000	0.000	0.000	0.000	0.000	0.000	0.000
1	13.009	0.397	75.125	0.146	0.351	0.191	0.516	0.003	1.641	0.849	0.650
2	12.549	0.781	139.573	0.568	0.743	0.389	2.269	0.005	6.380	3.183	3.027
3	12.105	1.150	194.581	1.232	1.086	0.562	5.131	0.007	13.840	6.663	6.986
4	11.677	1.507	241.250	2.106	1.384	0.712	8.963	0.009	23.676	10.999	12.327
5	11.264	1.851	280.560	3.162	1.643	0.842	13.636	0.011	35.572	15.947	18.868
6	10.866	2.183	313.382	4.375	1.866	0.955	19.036	0.012	49.249	21.302	26.447
7	10.482	2.503	340.491	5.722	2.057	1.051	25.061	0.013	64.452	26.895	34.918
8	10.111	2.812	362.576	7.183	2.219	1.132	31.617	0.014	80.954	32.588	44.148
9	0.000	2.812	324.828	8.632	2.097	1.060	38.250	0.013	97.361	37.651	53.551
10	0.000	2.817	290.217	9.954	1.916	0.968	44.331	0.012	112.356	41.492	62.199
11	0.000	2.817	259.310	11.158	1.751	0.885	49.886	0.011	126.055	44.292	70.099
12	0.000	2.817	231.709	12.255	1.600	0.808	54.962	0.010	138.573	46.214	77.317
24	0.000	2.817	60.377	19.794	0.551	0.278	90.681	0.003	226.665	37.001	128.108
36	0.000	2.817	15.942	22.134	0.195	0.099	103.120	0.001	257.354	18.823	145.794
48	0.000	2.817	4.288	22.695	0.071	0.036	107.574	0.000	268.345	8.326	152.126
60	0.000	2.817	1.184	22.637	0.027	0.014	109.227	0.000	272.426	3.475	154.474
72	0.000	2.817	0.340	22.363	0.011	0.006	109.879	0.000	274.035	1.429	155.401
84	0.000	2.817	0.105	22.014	0.006	0.003	110.168	0.000	274.752	0.606	155.812
96	0.000	2.817	0.037	21.642	0.003	0.002	110.324	0.000	275.154	0.283	156.038
108	0.000	2.817	0.017	21.266	0.003	0.001	110.431	0.000	275.369	0.158	156.171
120	0.000	2.817	0.011	20.891	0.002	0.001	110.538	0.000	275.584	0.110	156.278
132	0.000	2.817	0.009	20.526	0.002	0.001	110.659	0.000	275.798	0.091	156.385
144	0.000	2.817	0.008	20.164	0.002	0.001	110.659	0.000	276.013	0.082	156.493
156	0.000	2.817	0.007	19.809	0.002	0.001	110.712	0.000	276.227	0.079	156.600
168	0.000	2.817	0.007	19.460	0.002	0.001	110.766	0.000	276.442	0.076	156.707

[a]Based on long time lapse since first dermal contact until urine collection for subject B, whose observed urinary malathion acid metabolites accumulated up to 36 h were estimated to be 145.8 μg; all amounts, except for those remaining on the skin (mg) and evaporated to the air (mg), are expressed in micrograms; amounts in blood were negligible compared with those in other tissues and hence were not listed here.
[b]Under the assumption that the applied dose would be washed off 8 h after first dermal contact (see text for discussion on the use of applied dose).
[c]Includes both mono- and dicarboxylic acids.

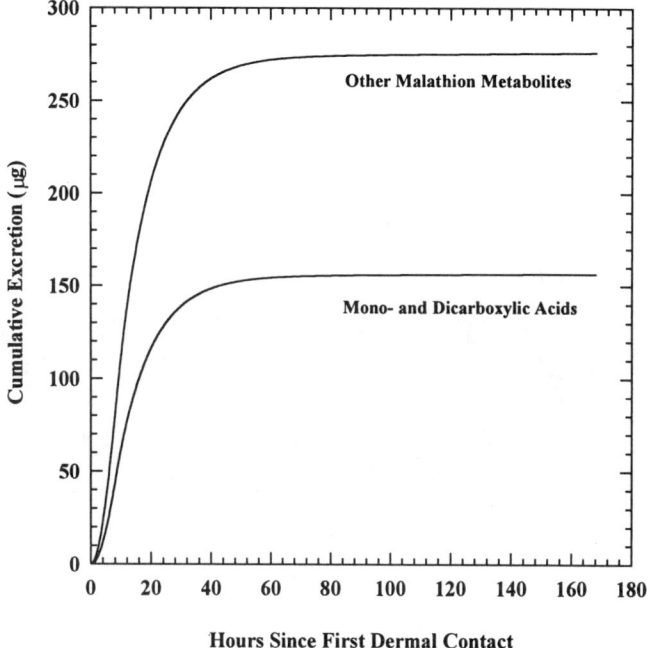

Figure 2. Excretion of urinary malathion metabolites (simulated for subject B).

in Figure 1, a chemical absorbed in the gastrointestinal tract is likely to be immediately available in the liver, the dominant site for chemical metabolism (including that for malathion). However, the same chemical from dermal exposure is likely to be more available to other tissue compartments and hence could be excreted without first being metabolized into MCA or DCA in the liver. Carboxylesterase is responsible for the degradation (deesterification) of malathion into the acid metabolites. It has been shown (35–37), however, that in the general population the level of circulating carboxylesterase available in the blood is negligible compared with that in the liver.

Table III also suggests that the excretion half-time of malathion was approximately 12 h after first dermal contact, which was assumed to have lasted 8 h. At 12 h, the total amount of excretion shown in Table III (i.e., that listed under feces, urine, and acid metabolites) was 0.27 mg, approximately half of the total absorbed dose estimated (for long-time lapse). This finding was consistent with that observed earlier in the study by Ross et al. (6), in which a range from 4 to 12 h was reported as the excretion half-time of malathion applied to the skin in a vehicle (which included protein bait). The excretion data in Table III indicate that at 48 h after first dermal contact, more than 75% of an absorbed dose would be recovered as urinary (acid and other) metabolites with another 20% recovered as fecal metabolites. The urinary

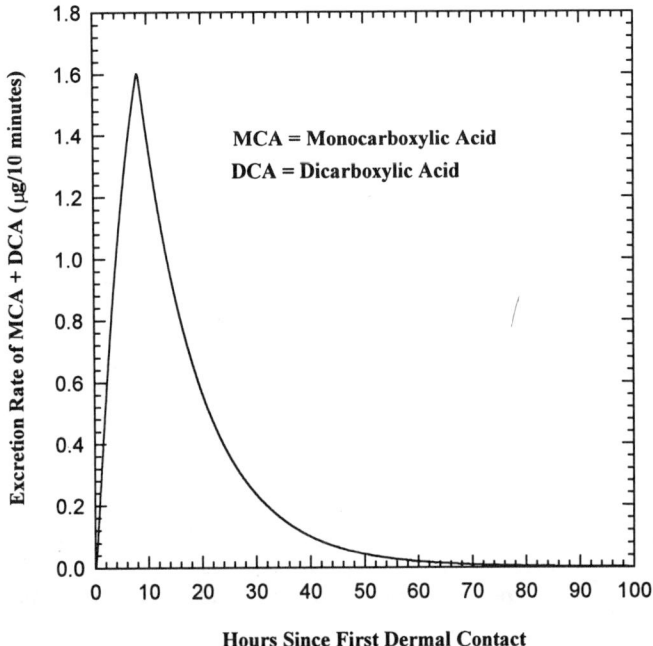

Figure 3. Excretion curve for malathion acid metabolites (simulated for subject B).

recovery simulated in this study was only 15–20% lower than those observed experimentally elsewhere (6, 38). The excretion rates simulated here for malathion acid metabolites (as shown in Figure 3) were similar to those estimated independently by Papanek and Woloshin (5) and those observed by others for total malathion metabolites (38, 39), thus further substantiating the kinetic equations used in the model.

Absorbed doses of malathion estimated here through PB-PK simulation were typically within an order of magnitude of the upper-bound estimates calculated earlier by Papanek and Woloshin (5). In that exposure assessment, relying on a different computational method and largely a different set of assumptions, Papanek and Woloshin also found the highest total absorbed doses (3.3 mg for short lapse and 9.6 mg for long lapse) in subject D. The lowest absorbed doses (30 μg for short lapse and 9 μg for long lapse) estimated in their exposure assessment also occurred in subject I. As shown in Table IV, the difference in time lapse tended to have an impact on the estimation of total absorbed dose.

In general, model predictions are greatly affected by the input parameters. Although many of the physiological and biochemical parameters applied in the PB-PK model were also used earlier by OEHHA (30, 31), it is

Table IV. Total Absorbed Dose of Malathion Estimated through Modeling

Subject	Case ID	Acid Metabolites[a] (μg/L Urine)	Malathion Dose (mg)[b]	
			Short Lapse	Long Lapse
Adult residents (70 kg)				
A	03	20.7	0.11	0.09
B	04	81.0	0.43	0.56
C	39	21.6	0.10	0.15
D	63	147.0	1.00	1.31
Agricultural workers (70 kg)				
E	AG49	45.0	0.22	0.20
F	AG52	81.0	0.40	0.36
G	AG53	40.0	0.20	0.18
Younger children (14 kg)				
H	48	225.0	0.30	0.30
I	61	24.9	0.03	0.03
Older children (35 kg)				
J	31	75.0	0.20	0.19
K	60	156.0	0.42[c]	0.40[c]

[a] The amount of total acid metabolites accumulated up to the assumed time lapse between first dermal contact and urine collection (see Table I) was used to simulate each participant's total absorbed dose of malathion; the daily urine output volumes for adults, younger children, and older children were assumed to be 1200, 500, and 800 mL, respectively (32).

[b] Each participant's dose was also simulated by using the shortest as well as the longest assumed number of hours between first dermal contact and urine collection; for all study cases, except subjects B, C, and D, the shortest time lapse was assumed to be 3 h (see footnotes in Table I for assumed time lapses).

[c] May be substantially overestimated because of overestimation of the child's body weight (32); as shown in Table I, the child was 5 years old at the time of the biomonitoring study.

conceivable that the input parameters listed in Table II might be best defined by their ranges, rather than by their means or mean-based point estimates. A more reliable and more complex approach to PB-PK modeling would thus be to use parameter values randomly sampled from their range, while taking into account their known or presumed statistical distribution. Many investigators have begun to utilize this approach to refining or improving their PB-PK models in risk assessment (14–16, 23, 24). For this approach, Monte Carlo simulations would also be required to obtain a representative sample of model predictions, which by definition would be statistical distributions rather than single points.

The Monte Carlo technique was not incorporated into the PB-PK modeling here, however, despite its increasing popularity. A major reason for this omission was that other parameters used in the model had even greater uncertainty. These included the assumed time lapse and the estimated total acid metabolites estimated for the 11 study cases. As shown in Table IV, dose simulation was affected somewhat by the assumed time lapse. A great deal of uncertainty was associated with this time variable, because its measurement was primarily based on the participant's recollection. The values estimated for the observed total acid metabolites were also subject to substantial uncertainty, in that direct measurements were not made on MCA and total urine outputs were not collected for the participants. Because the purpose of this case study was illustrative and the biomonitoring study had limitations in its experimental design, the complex Monte Carlo approach to PB-PK modeling was not warranted here. (For a limited discussion of improved experimental design see the Conclusion section.)

Although the Monte Carlo approach was not used in this case study, over 70 additional simulations were made to determine the sensitivity of dose estimation to the key biochemical parameters listed in Table II (for all subjects). When the partition coefficient value for any of the listed tissues other than skin was deliberately doubled, the absorbed dose predicted was altered by about 15% or less. Even when either the V_{max} or K_M value was doubled, the dose was altered by no more than 35%. In this sensitivity analysis, model predictions were found to be most affected (up to as much as two-fold) by the skin–blood partition coefficient. Such a finding was not unexpected, however, because the tissue–blood partition coefficient is defined as the concentration of a chemical in the tissue compared with that in the blood at steady state. If more malathion were bound in the skin (as would be the case with an increased partition coefficient), less would be available to the liver for metabolism. This effect, in turn, would require an additional dose absorbed into the body (the skin) so that the same theoretical amount of malathion could be available to the liver for metabolism.

Several input parameters listed in Table II were assigned arbitrary values. These parameters included the constants for skin permeability and for evaporation into air. In this case study, the total absorbed dose (not the applied dermal dose) was of interest. The PB-PK model was constructed for a dose applied to a skin surface area of $1\ cm^2$. The applied dose was at first assigned an arbitrary value for a given study case and then was varied in subsequent runs until the predicted accumulation of acid metabolites matched that estimated from the biomonitoring data. In short, the model was independent of the total exposed surface area. If the actual exposed surface area were 1000 cm^2, the model dose corresponding to a 1-cm^2 surface area would need to be reduced by 1000-fold (so that the total dermal dose that would penetrate through the skin would remain the same). This proposition is in accord with the following equation given by McDougal et al. (21) and by Knaak et al. (25):

$$ABS = P \times A \times C \times T$$

where P is the skin permeability constant, A is the surface area exposed, C is the exposure concentration, T is the permeation time, and ABS is the total amount of chemical absorbed up to permeation time T. An extension of this equation made here is that ABS is quantifiable by summing the amounts of chemical simulated for all the internal tissue compartments involved. The amount of chemical in each compartment is estimated by the postulated kinetic equations so that the total comprises the ABS. Although the amounts of chemical in the internal tissues change over time, their sum after permeation stops (in this case after 8 h) always remains the same; this is because the chemical (or its metabolites) has to be present in some of the internal tissue compartments. This assurance of additivity is based on the commonly encountered biochemical principle known as the mass-balance theory.

Unlike most air levels of organic vapors assumed for dermal exposure assessment, the total applied malathion dose (i.e., $A \times C$) received through dermal contact by each subject in this case study was expected to dissipate over time. As reflected in equation 2, its quantity on the skin diminishes over time because of permeation and evaporation. Insofar as the constants for skin permeability and evaporation are comparatively small and their ratio does not alter substantially, the applied dose that remains on the skin will dissipate at a relatively very low, fairly steady rate (for an example, see Table III). Consequently, the absorbed dose simulated here should not be affected by the constants assumed for skin permeability or for evaporation, unless these assigned values, particularly that for the latter constant, are far off the range limits proposed by OEHHA (30, 31).

Conclusion

As discussed in several earlier chapters in this volume, classification of exposure groups is a cornerstone of well-designed studies in environmental epidemiology. The increasing sensitivity and sophistication of analytical tests have greatly expanded the array of biomarkers available to the epidemiologist. Data on biomarkers in biological fluids, regardless of how accurate, usually do not correspond in any direct way to the dose absorbed. As demonstrated in this chapter, PB-PK models provide an essential interpretive tool when applied to biomarker data.

PB-PK modeling was used in this case study to estimate the total absorbed dose of malathion in seven adults and four children. The subjects were alleged to have come into contact with the Medfly eradication aerial sprays that were applied over portions of southern California during the fall of 1989 and continuing through July 1990. Results predicted through this PB-PK simulation suggest that for the adults (70 kg), the highest absorbed dose of malathion was 1.3 mg in the spray event studied. Among the younger (14 kg) and the older (35 kg) children, the highest estimated total doses were 0.3 mg and 0.4 mg, respectively. These estimations were based on a conservative approach, in that the model assumed an above-average daily urine output for the study cases and considered the shortest as well as the longest probable time lapse from first dermal contact until urine collection. In addition, the model assumed that as much as 20% of an absorbed dose of malathion would be excreted in feces, thus resulting in a lower production of the acid metabolites that would lead to a higher quantity of the *ABS* simulated.

In addition to its utility in estimating the total absorbed dose of a chemical, PB-PK simulation can be used to estimate simultaneously the various time-dependent tissue concentrations. The amounts of malathion or its metabolites simulated for the various internal tissues shown in Table III, when divided by their corresponding tissue volume, would provide the tissue concentrations of interest. This type of pharmacokinetic information is crucial for health risk assessment if the toxic end point in question is organ-specific. Gearhart et al. (*18*) recently demonstrated the use of PB-PK models for predicting cholinesterase inhibition in rats. Although malathion can cause acute cholinergic reactions in humans at high dosages, the PB-PK model used in this case study was not revised to separate brain from the vessel-rich groups. This is because the objective here was not to estimate the various tissue concentrations per se but rather to estimate total dosage.

Future studies need experimental design considerations to exploit more fully the sophistication and precision of PB-PK models. For example, analytical methods are needed to quantify the breadth of known malathion urinary metabolites, not just DCA or the sum of DCA and MCA. The PB-PK simulations performed would have been more reliable had the theoretical accumulation of malathion urinary metabolites (for each study case) been fit simultaneously as well as statistically to a set of experimental values corresponding to definite time points, such as at 6, 12, 24, and 48 h. Finally, additional information is needed on the background exposures to malathion in the general population. Confounding due to dietary exposures was not evaluated, other than by a questionnaire. Nevertheless, this pilot study was believed to be useful in providing a rough estimate of the total absorbed dose of malathion for individuals allegedly exposed to the Medfly eradication aerial sprays, particularly in light of the fact that the subjects were recruited largely from the general population in sprayed areas and that recruitment was based primarily on self-reported, alleged exposures surfacing in complaints to local health authorities.

References

1. Gallo, M. A.; Lawryk, N. J. In *Handbook of Pesticide Toxicology;* Hayes, W. J.; Laws, E. R., Eds.; Academic: Orlando, FL, 1991; Vol. 2, Chapter 16.
2. Bradway, D. E.; Shafik, T. M. *J. Agric. Food Chem.* **1977**, *25*, 1342–1344.
3. Fenske, R. A.; Leffingwell, J. T. *J. Agric. Food Chem.* **1989**, *37*, 995–998.
4. Draper, W. M.; Wijekoon, D.; Stephens, R. D. *J. Agric. Food Chem.* **1991**, *39*, 1796–1801.
5. Papanek, P. J., Jr.; Woloshin, K. A. *Evaluation of Malathion Urinary Metabolites in Individuals Potentially Exposed to Aerial Applications of Malathion Bait in the Medfly Eradication Campaign;* Los Angeles County Department of Health Services: Los Angeles, CA, 1991; Technical Report, Toxics Epidemiology Program.
6. Ross, J. H.; Thongsinthusak, T.; Krieger, R. I.; Frederickson, S.; Fong, H. R.; Taylor, S.; Begum, S.; Dong, M. H. *The Toxicologist* **1991**, *11*, 160.
7. Gerlowski, L. E.; Jain, R. K. *J. Pharm. Sci.* **1983**, *72*, 1103–1127.
8. Ramsey, J. R.; Andersen, M. E. *Toxicol. Appl. Pharmacol.* **1984**, *73*, 159–175.
9. McDougal, J. N.; Jepson, G. W.; Clewell, H. J., III; MacNaughton, M. G.; Andersen, M. E. *Toxicol. Appl. Pharmacol.* **1986**, *85*, 286–294.
10. Andersen, M. E.; Clewell, H. J., III; Gargas, M. L.; Smith, F. A.; Reitz, R. H. *Toxicol. Appl. Pharmacol.* **1987**, *87*, 185–205.
11. *Pharmacokinetics in Risk Assessment—Drinking Water and Health;* National Research Council, Ed.; National Academy of Science: Washington, DC, 1987; Vol. 8.
12. D'Souza, R. W.; Andersen, M. E. *Toxicol. Appl. Pharmacol.* **1988**, *95*, 230–240.
13. Reitz, R. H.; McDougal, J. N.; Himmelstein, M. W.; Nolan, R. J.; Schumann, A. M. *Toxicol. Appl. Pharmacol.* **1988**, *95*, 185–199.
14. Farrar, D.; Allen, B.; Crump, K.; Shipp, A. *Toxicol. Lett.* **1989**, *49*, 371–385.
15. Portier, C. J.; Kaplan, N. L. *Fund. Appl. Toxicol.* **1989**, *13*, 533–544.
16. Bois, F. Y.; Zeise, L.; Tozer, T. N. *Toxicol. Appl. Pharmacol.* **1990**, *102*, 300–315.
17. Corley, R. A.; Mendrala, A. L.; Smith, F. A.; Staats, D. A.; Gargas, M. L.; Conolly, R. B.; Andersen, M. E.; Reitz, R. H. *Toxicol. Appl. Pharmacol.* **1990**, *103*, 512–527.
18. Gearhart, J. M.; Jepson, G. W.; Clewell, H. J., III; Andersen, M. E.; Conolly, R. B. *Toxicol. Appl. Pharmacol.* **1990**, *106*, 295–310.
19. Ishida, S.; Sakiya, Y.; Ichikawa, T.; Taira, Z.; Awazu, S. *Chem. Pharm. Bull.* **1990**, *38*, 212–218.
20. Knaak, J. B.; Al-Bayati, M. A.; Raabe, O. T.; Blancato, J. N. In *Prediction of Percutaneous Penetration—Methods, Measurements, and Modeling;* Scott, R. C.; Guy, R. H.; Hadgraft, J., Eds.; IBC Technical: London, 1990; pp 1–18.
21. McDougal, J. N.; Jepson, G. W.; Clewell, H. J., III; Gargas, M. L.; Andersen, M. E. *Fund. Appl. Toxicol.* **1990**, *14*, 299–308.
22. Reitz, R. H.; McCroskey, P. S.; Park, C. N.; Andersen, M. E.; Gargas, M. L. *Toxicol. Appl. Pharmacol.* **1990**, *105*, 37–54.
23. Bois, F. Y.; Woodruff, T. J.; Spear, R. C. *Toxicol. Appl. Pharmacol.* **1991**, *110*, 79–88.
24. Bois, F. Y.; Paxman, D. G. *Regul. Toxicol. Pharmacol.* **1992**, *15*, 122–136.
25. Knaak, J. B.; Al-Bayati, M. A.; Raabe, O. G. In *Health Risk Assessment—Dermal and Inhalation Exposure and Absorption of Toxicants;* Wang, R. G. M.; Knaak, J. B.; Maibach, H. I., Eds.; CRC: Boca Raton, FL, 1993; pp 3–29.

26. Bischoff, K. B. In *Chemical Engineering in Medicine and Biology;* Hershey, D., Ed.; Plenum: New York, 1967; pp 417–446.
27. Gibaldi, M.; Perrier, D. *Pharmacokinetics;* Marcel Dekker: New York, 1982, Chapter 9.
28. Andersen, M. E. *Ann. Occup. Hyg.* **1991**, *35*, 309–321.
29. Dong, M. H. *Microcomputer Programs for Physiologically-Based Pharmacokinetic (PB-PK) Modeling;* Worker Health and Safety Branch, California Department of Pesticide Regulation: Sacramento, CA, 1991; Technical Report HS–1635.
30. *Health Risk Assessment of Aerial Application of Malathion-Bait;* California Environmental Protection Agency, Office of Environmental Health Hazard Assessment: Sacramento, CA, 1991. (Reprints available from Copies Unlimited, 5904 Sunset Boulevard, Los Angeles, CA 90028.)
31. Rabovsky, J.; Brown, J. P. *Intern. J. Occup. Med. Toxicol.* **1993**, *2*, 131–168.
32. *Scientific Tables;* Diem, K.; Lentner, C., Eds.; Ciba-Geigy: New York, 1973; p 661.
33. Bingham, S. A.; Cummings, J. H. *Hum. Nutr. Clin. Nutr.* **1985**, *39*, 343–353.
34. Greenberg, G. N.; Levine, R. J. *J. Occup. Med.* **1989**, *31*, 832–838.
35. Main, A. R.; Braid, P. E. *Biochem. J.* **1962**, *84*, 255–263.
36. Mendoza, C. E.; Shields, J. B.; Greenleaf, R. *Comp. Biochem. Physiol.* **1977**, *56C*, 189–191.
37. Talcott, R. E.; Pond, S. M.; Ketterman, A. J.; Beker, C. E. *Toxicol. Appl. Pharmacol.* **1982**, *65*, 69–74.
38. Feldmann, R. J.; Maibach, H. I. *Toxicol. Appl. Pharmacol.* **1974**, *28*, 126–132.
39. Wester, R. C.; Maibach, H. I.; Bucks, D. A. W.; Guy, R. H. *Toxicol. Appl. Pharmacol.* **1983**, *68*, 116–119.

RECEIVED for review November 2, 1992. ACCEPTED revised manuscript January 28, 1993.

15

An Epidemiological Assessment of the Cantara Metam Sodium Spill

Acute Health Effects and Methyl Isothiocyanate Exposure

Richard A. Kreutzer[1], David J. Hewitt[1,3], and William M. Draper[2]

[1]Environmental Epidemiology Section, California Department of Health Services, 5900 Hollis Street, Suite E, Emeryville, CA 94608
[2]Hazardous Materials Laboratory, California Department of Health Services, 2151 Berkeley Way, Berkeley, CA 94704

> *This chapter describes an environmental epidemiology case study involving acute health effects of the pesticide metam sodium released in a railroad transportation accident in northern California. A total of 23,500 kg of metam sodium spilled into the Sacramento River, destroying nearly all aquatic life in a 45-mi stretch of river and exposing 2129 residents of Dunsmuir to metam sodium's volatile decomposition products, which include the potent irritant methyl isothiocyanate (MITC). Reported health effects, including nonspecific neurologic complaints (headache and dizziness) and irritation (eye, respiratory tract, gastrointestinal tract, and skin) were consistent with MITC exposure. In Dunsmuir, 14% of residents sought medical attention. The case study demonstrates how health effects in a 1-month period following the spill were ascertained by using questionnaires, pesticide illness reports from private physicians, and hospital inpatient and emergency room medical records. An environmental fate and transport model suggests that the highest air concentrations of MITC in Dunsmuir were less than 160 ppb and may have occurred over 12 h after the spill as a result of a rapid photochemical conversion of metam sodium to MITC in sunlight.*

The Cantara Metam Sodium Spill

The Sacramento River in northern California has long been renowned for its beauty and excellent trout fishing. Originating near Mt. Shasta, it flows

---

[3]Current address: 13500 Cheral Parkway, Apartment 790D, Little Rock, AK 72211.

south to Lake Shasta (Figure 1). Human population along this segment of the river is generally sparse. On Sunday, July 14, 1991, at 21:40 h, a Southern Pacific Railroad train consisting of 97 cars and four locomotives derailed on a tight bridge curve spanning the Sacramento River, known as the "Cantara Loop", located 6 mi north of Dunsmuir. A locomotive, six empty freight cars, and an unlabeled tank car left the tracks. The tank car came to rest in a partially inverted position in the shallows directly beneath the bridge (1).

Crew members reported odors coming from the derailed cars shortly after the derailment. At 04:30 on July 15, railroad officials reported a nonleaking puncture in the derailed tank car above the water line. It was estimated that approximately 1500 gal of metam sodium might have leaked into the river. As reports of health effects and dead fish downstream became

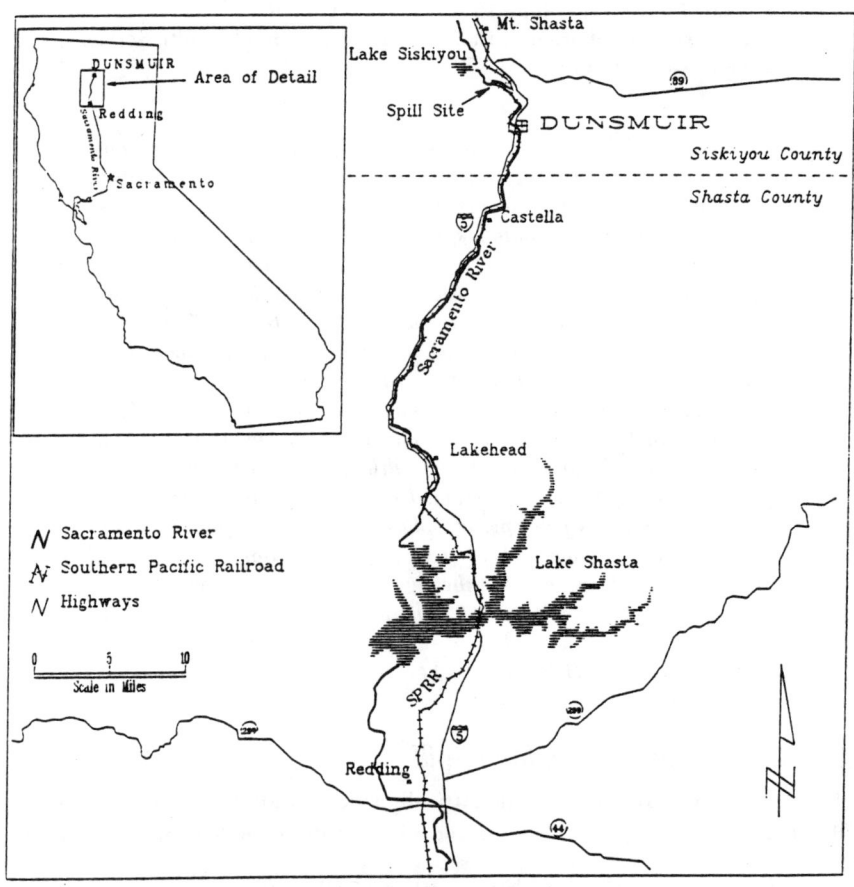

Figure 1. Map of affected area.

available later that morning, the extent of the spill was reevaluated. Two additional submerged holes in the car were found where it was resting on the river bed. By approximately noon on July 15, it was estimated that nearly all of the 19,500 gal of pesticide transported in the tank car were released, probably within 1 h of the derailment (1).

Early Reports of Public Health and Environmental Effects

The plume traveled en masse southward at approximately 1 mph until the early hours of July 17 when it reached Lake Shasta, 45 mi from the spill site. Aquatic life along this section of the river was severely affected. The California Department of Fish and Game (CDFG) reported fish kills in the hundreds of thousands.

The largest population center on the affected segment of the river is the town of Dunsmuir (population 2129, 1990 census) (2) (Figure 1). Several much smaller communities (Castella, Sweetbrier, La Moine, Pollard Flats, Delta, and Lakehead) and campgrounds are located farther downstream.

Within 8 h of the spill, the county health department had received from Dunsmuir residents numerous complaints of irritated eyes and unpleasant odors. The tail end of a light yellow-green plume was reported to pass through Dunsmuir around 07:45 (1). The Mt. Shasta hospital, located a few miles upstream from the spill site, began seeing patients with exposure-related illnesses in its emergency room by 11:00 on July 15. A temporary shelter for voluntary evacuees was established at the Dunsmuir High School on the afternoon of July 15 and remained open until 20:00 on July 20. Paramedics were stationed at the shelter to assess, treat, and triage health complaints.

Metam Sodium Chemistry and Properties

Metam sodium (sodium methyldithiocarbamate or sodium methylcarbamodithioate) was patented in 1958 as a soil fumigant and nematicide (3). At high concentrations, metam sodium also is an herbicide, and it is widely used as a wood preservative. Agricultural usage of metam sodium has increased substantially in recent years, because other previously used fumigants such as 1,3-dichloropropene were removed from the market due to health concerns (4).

The fumigant activity of metam sodium is attributed to its volatile breakdown product, methyl isothiocyanate (MITC)—metam sodium itself is an involatile salt. MITC also is a registered pesticide that has a spectrum of activity similar to that of metam sodium. Surprisingly, alkyl isothiocyanates, including MITC, occur naturally in plants, but they are usually sequestered as glucosinolates.

Metam sodium is commonly formulated as a 32.7%-by-weight aqueous solution, the mixture that was spilled into the Sacramento River. It is prepared commercially by reacting carbon disulfide, methylamine, and sodium hydroxide. At high concentrations, metam sodium solutions remain slightly basic, and the pesticide is relatively stable. At very high and low pH values, metam sodium decomposes via different pathways (Figure 2). Below pH 4.0, metam sodium [pK_a = 2.89 (5)] is protonated, and free methyldithiocarbamic acid rapidly converts back to the starting materials, methylamine and carbon disulfide. At high pH values (>11), metam sodium loses a second proton, and the dianion decomposes to MITC and sulfide.

Typically, the concentrated metam sodium formulations are diluted with water and are applied directly to soil. Under these conditions, metam sodium is rapidly converted in high yield to MITC (6).

The fate of metam sodium in natural water is not as well-understood. The acid- and base-catalyzed decomposition reactions (Figure 2) are not important at pH 7.0, where metam sodium occurs as its conjugate base [CH_3NH—$C(S)S^-$]. Dissolved oxygen may oxidize metam sodium to a disulfide, but this oxidation is not thermodynamically favored below pH 9.5 (7). At near-neutral or slightly acidic pH, metam sodium decomposes via a monomolecular cleavage reaction, giving rise to MITC and hydrogen sulfide (Figure 2). The cleavage reaction is believed to involve a planar intermediate in which a hydrogen is bonded to both sulfur and nitrogen. In addition to decomposition by cleavage, metam sodium is subject to photochemical decomposition in natural water. Metam sodium is estimated to have a half-life of between 30 min and 1.6 h for dilute solutions irradiated in summer sunlight (8, 9). The photodecomposition products of metam sodium include MITC, which accounts for 21–24 mol% of metam sodium reacting (8).

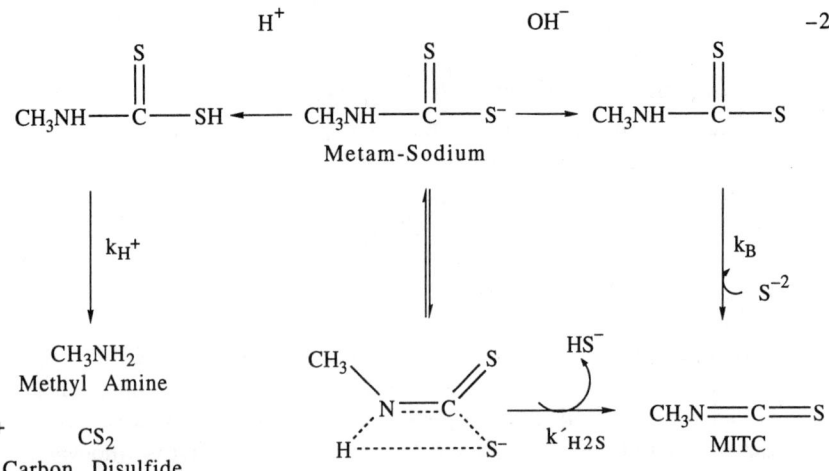

Figure 2. *Metam-sodium decomposition pathways at acidic ($k_H{}^+$), basic (k_B) and near-neutral (k'_{H_2S}) conditions.*

Toxicology of Metam Sodium and Decomposition Products

Acute Health Effects. Because of its involatility, significant human exposure to metam sodium was not expected. Exposure to its volatile breakdown products, which could be released into the atmosphere and dispersed over a large area, was of greater concern. Both MITC and hydrogen sulfide, the main breakdown products, are chemical irritants. Low concentrations of H_2S (e.g., 10–200 ppm) can cause localized eye and respiratory-tract irritation. Higher concentrations may produce systemic symptoms, such as nausea, vomiting, diarrhea, headache, dizziness, confusion, weakness, tachypnea, tachycardia, cardiac arrhythmia, and sweating (10).

Less is known about the health effects of MITC. Also known as "methyl mustard oil", MITC [dose that is lethal to 50% of test subjects (LD_{50})]: 97 mg/kg rat, oral, acute) is considerably more toxic than metam sodium (LD_{50}: 820–1800 mg/kg rat, oral, acute) and was once considered as a chemical warfare agent because of its irritant properties. MITC is known to be extremely irritating to skin and mucous membranes and at much lower concentrations than H_2S (10–12). Although other acute health effects of MITC in humans are less well described, the health effects would appear to be similar to those seen with H_2S exposure, on the basis of limited data from animal studies and case reports of occupational exposure (13, 14).

An important difference between MITC and H_2S is the relationship between their odor and irritant thresholds. MITC is an insidious poison capable of inducing symptoms at concentrations below its odor threshold [odor threshold 100 ppb; irritant threshold 70 ppb (12)]. H_2S, in contrast, can be sensed at concentrations well below those producing irritant symptoms [odor threshold 1–2 ppb; irritant threshold 10,000 ppb (10)].

Chronic Health Effects. A number of laboratory animal studies normally required by the U.S. Environmental Protection Agency (U.S. EPA) for pesticide registration have not yet been completed. Therefore, specified data on chronic, carcinogenic, and reproductive effects are incomplete (4).

Metam sodium has not been examined in a chronic cancer study (4, 13), although MITC cancer bioassays in rats and mice have had negative results. Metam sodium is a teratogen in some experimental animals and produces neural-tube defects in rats and rabbits at high doses (4). At levels that showed minimal maternal toxicity, MITC administration was associated with fetal growth retardation in rabbits. Effects on humans are unknown.

Individuals chronically exposed to irritant gases similar to MITC occasionally develop prolonged bronchial hyperresponsiveness (i.e., reactive airway dysfunction syndrome) following an initial chemical bronchitis (15).

Epidemiological Investigation Objectives

Substantial unknown factors complicated predicting the potential health effects of the spill. The rail car was not placarded as carrying a hazardous material, and identification and notification were thus delayed. Even after the contents of the car were identified as metam sodium, the composition of the formulation, its contaminants, and breakdown products remained obscure. Although previous chemical and occupational health studies provided an indication of the probable environmental fate of metam sodium and potential health effects of MITC, the behavior of a large quantity of the pesticide in the Sacramento River ecosystem could not be predicted with certainty.

Environmental data on metam sodium or MITC levels were not available for several days after the spill. Metam sodium is difficult to analyze directly, and can be determined only by high-pressure liquid chromatography (HPLC) with an unconventional micellar mobile phase (16). Metam sodium is usually determined indirectly by measuring its decomposition product, MITC. Before the spill, California environmental testing laboratories were inexperienced in analyzing either metam sodium or MITC. Although water samples were taken within 14 h of the accident, laboratories had neither analytical reference standards nor validated analytical methods.

Technical difficulties were also experienced in obtaining air data. The first samples, taken within 24 h of the spill, used an adsorbent (XAD-2) with trapping efficiency inadequate for determination of MITC. The earliest reliable air data on MITC correspond to 48 h after the spill, when an appropriate trapping medium (charcoal) was substituted. Samples had to be flown to a laboratory over 200 mi away, which further delayed analysis.

Thus, the first objective of the epidemiological investigation was to determine quickly, in the absence of environmental data, the immediate risk to public health and the extent of health effects from the spill. In brief, telephone calls and visits to area hospitals and the shelter were made from July 15–20. Health-care providers at these sites were interviewed about the numbers of spill-related patients evaluated and the types and severity of symptoms. Patients' medical charts were reviewed, and reported symptoms were assessed for consistency with known toxicological end points of metam sodium or for unexplained symptomatology. This preliminary emergency assessment indicated that (1) a substantial number of individuals reported symptoms from the spill, most of which were not severe enough to require hospitalization; (2) there were no fatalities as a result of the spill; and (3) reported symptoms were generally consistent with exposure to irritant vapors such as MITC.

A second objective of the investigation was formal documentation and analysis of spill-related health effects reported within the first month after the incident. To achieve this objective, various methods to assess exposure and to ascertain cases were explored. Information from such a study could be used to define more clearly the human toxicology of metam sodium and its breakdown products and to document more fully the acute health impact of the incident. Probable MITC exposure situations were constructed on the basis of a metam sodium environmental fate model and correlated with the observed health effects. Methods and findings of this analysis are described further on.

Methods

Case Ascertainment and Analysis. We examined spill-related health effects reported from July 14 through August 16, 1991, from the temporary evacuation and triage shelter in Dunsmuir, private physicians in the Dunsmuir–Mt. Shasta region, and in area hospitals. These data sources are described further on in more detail. For each information source, "cases" were defined as persons reporting symptoms that they believed were due to the spill during the period of study. Exposure was necessarily based on self-report. Cases potentially could include area residents, tourists, or individuals traveling through the area who sought medical evaluation.

Data Sources. Dunsmuir shelter: A one-page, self-administered questionnaire for patients was designed by California Department of Health Services (CDHS) staff and distributed to the shelter on July 15. In addition to name, gender, age, address, and date completed, the questionnaire included a list of possible symptoms, based on effects of probable metam sodium decomposition products, which could be checked or circled. Information on other symptoms, odors smelled, and location, time, and duration of exposure also were requested.

Individuals (or parents of young children) using the Dunsmuir shelter medical facility during its period of operation from July 15–20 completed the questionnaires, which were supplied and collected by shelter staff. In addition, investigators were permitted to review a separate card file that shelter paramedics had maintained for all patient contacts. Questionnaires were supplemented with paramedic data, when possible, to complete the case profile.

Local Private Physicians. Pesticide illness reports (PIRs) were a second major data source. These are one-page forms that must be completed by any California physician who examines a patient with health problems suspected to be pesticide-related. The forms are submitted to both the county and state health authorities. Items on the form include the patient's name, gender, age, and address; date of evaluation; and date, time, location, and route of exposure. Physicians also are asked to describe briefly symptoms, physical findings, and treatment and to indicate whether the patient was pregnant at the time of exposure. Copies of all spill-related PIRs for the study period that had been received by Siskiyou County Environmental Health Department through September 1991 were obtained.

Area Hospitals. Three hospitals (a small community hospital in Mt. Shasta and two larger regional hospitals in Redding) within 25 mi of the spill area were identified and were presumed most likely to have seen spill-related illnesses. On July 15 and 16, investigators alerted these hospitals of the CDHS study and asked them to participate. CDHS requested access to all emergency room and inpatient medical records of patients identified by hospital staff as spill-related. Beginning on July 15, the Mt. Shasta hospital staff also agreed to distribute and collect questionnaires identical to those distributed at the shelter. Patients voluntarily completed the questionnaire while waiting for evaluation. If completed, questionnaires were included in the patient's permanent medical record and were used by abstractors to supplement information provided by the attending physician.

After the CDHS field investigation team left, the Mt. Shasta hospital was contacted daily by telephone through August 16 to monitor the number of spill-related patients seen. Because the two hospitals in Redding reported very few spill-related patient contacts during the first week after the spill, they were not surveyed after July 19. The remaining spill-related patient charts for the study period from the Mt. Shasta hospital were abstracted on-site in August.

A single abstraction form was designed to extract data from the questionnaires, PIRs, and hospital patient records. The abstraction form was prepared in an Epi Info (17) computer database for direct key entry of records, when possible. Epi Info also provides for convenient data management and rapid preliminary analysis.

Data Analysis. Once records were entered into the database, they were reviewed for duplicates, because some records were abstracted from multiple sources. If the recorded visits for the same person occurred on different days, they were considered separate records, regardless of the reporting source. If the visits to different reporting sources occurred on the same day, it was usually the result of paramedics at the shelter triaging patients to either the local physicians' offices or to the Mt. Shasta emergency room, thus, these visits were recorded as a single patient contact. Discrepancies in the information provided by the different sources were resolved where the correct information was clear (e.g., key entry error). Where resolution was not possible, the emergency room or private physician record was accepted over the shelter record.

Data were analyzed by date and site of evaluation, age, gender, and city of residence. Frequencies of reported odors and symptoms were calculated. Symptoms were stratified by age, sex, evaluation site, evaluation date, odor detection, and smoking history. A one-way analysis of variance or student's t-test was used to determine whether the average number of symptoms reported per case differed for any of the stratified variables.

All hospitalizations and pregnancies identified through the surveillance were reviewed. In October investigators conducted a telephone follow-up survey of women pregnant at the time of the spill. Questions on past pregnancy history, number of weeks pregnant at the time of the spill, progress of pregnancy since the spill, and pregnancy outcome were included.

Estimating MITC Exposures in Dunsmuir. Authorities evaluating the effects of the pesticide spill needed both qualitative and quantitative information on chemical exposures. The available information indicated that MITC was the principal transformation product. In view of its toxicity (relative to hydrogen sulfide) and its potential to volatilize rapidly from water, MITC was presumed to pose the greatest threat to public health. Thus, estimating MITC exposures in the community of Dunsmuir was a high priority because of the town's close proximity to the spill and its comparatively large population clustered along the river. The leading edge of the plume reached Dunsmuir early in the morning of July 15, when most residents would have been at home.

Peak MITC exposures were estimated by using a metam sodium environmental fate and transport model coupled to an atmospheric dispersion model described in detail elsewhere (24). The MITC source strength was based on estimations of (1) the initial water concentration of metam sodium, (2) the rate of conversion of metam sodium to MITC in darkness and daylight, and (3) the rate of MITC volatilization from the river's surface. The dilution of airborne MITC between the source and the receptor community was approximated by using a Gaussian plume atmospheric dispersion model, which assumed an upwind, virtual point-source as described by the U.S. EPA (7, 18). This technique is widely used in permitting hazardous-waste management facilities where fenceline concentrations of toxic substances are of interest.

The emission rate of MITC from the river's surface was determined by two consecutive first-order processes: cleavage yielding MITC followed by MITC volatilization from water. Rate constants for metam sodium cleavage and MITC yield were determined in the dark and in simulated sunlight in laboratory studies described elsewhere (21). Based on the laboratory experiments, metam sodium was expected to decompose with a rate constant of 1.7×10^{-4} min^{-1} in the 15 °C Sacramento River water. The experimental yield of MITC was 15 mol%, meaning that each kilogram of metam sodium decomposing in the dark yielded 85 g of MITC.

In daylight metam sodium degraded much more rapidly. The rate of photodecomposition of metam sodium in sunlight was approximated by using models again developed by the U.S. EPA (19, 20). These models accurately predict direct photolysis rates of pollutants with corrections for season, latitude, water depth, and other parameters. The absorption spectrum of metam sodium and the photochemical quantum yield also are required inputs to the model and were determined experimentally by Draper and Wakeham (21). The Φ_{solar} determined for metam sodium was 0.4 ± 0.1, exceptionally high for any photolabile compound. The observed yield of

MITC from the photochemical reaction was 19%, similar to that reported previously (8). The U.S. EPA model predicted a half-life of 2.5 h for metam sodium in midday summer sunlight with a water depth of 0.4 m.

Environmental models require various assumptions and simplifications, which substitute for missing information and provide easily solved mathematical expressions. The characteristics of the model pertaining to the spill (e.g., plume dimensions, current and wind speeds, and daylight–darkness regimen) have been outlined in detail elsewhere (24).

Geographic Distribution of Cases. A map of the Dunsmuir area showing the location of city limits, the major streets and highways, and the Sacramento River, was produced with the computer program ARC/Info (22). Home addresses of cases listing a Dunsmuir residence were determined from database information. For cases in which no street address was given (i.e., address missing or post office box given), a 1991 telephone directory of the area was consulted. As a final method of determining home addresses, the Dunsmuir Post Office was contacted. Cases in which the street address could not be located were excluded. Case residences then were geographically coded and plotted on the Dunsmuir map.

Rates of illness were calculated for the Dunsmuir population and stratified by age and gender. Cases living in the city constituted the numerator, and the 1990 Dunsmuir census population was the denominator.

If the major component of MITC exposure occurred as the plume passed through the town, as was initially assumed, both exposures (and symptom attack rates) would be expected to decline with distance from the river. Serial 300-ft concentric regions or zones around the river were constructed. Because of the irregular city boundaries, the zones included some areas outside the city limits. Zones also were extended approximately ½ mi past the southern city boundary to include cases reporting a Dunsmuir residence but living just south of the city limits. Populations of each zone were determined from census block data. Where a block crossed two or more zones, the proportion of the block population in the zone was presumed equal to the proportion of the block's housing units located within that zone. Rates of illness were calculated and contrasted for each zone.

Results

Data Sources. Information was obtained for 848 spill-related medical visits occurring between July 15 and August 16, 1991. A total of 705 separate individuals with 115 seeking medical attention more than once (89 seen twice, 24 seen three times, and 2 seen four times) were visited. The number of medical evaluations was approximately equally distributed among the triage center (338), physicians' offices (223), and emergency rooms (278). Of the 278 emergency-room visits recorded, 257 (92.4%) occurred at the Mt. Shasta hospital, and 21 (7.6%) occurred at the two Redding hospitals. Information from 223 visits to private physicians in the area was primarily obtained from PIRs received from 17 different providers. A total of 23 individuals evaluated at the shelter sought additional evaluation from emergency rooms or private physicians.

Characteristics of Cases. The majority of individuals who sought medical attention were from Dunsmuir (70.6%), Mt. Shasta (7.2%), and Castella (6.4%), the communities closest to the spill site. The age and gender distributions of those reporting symptoms are shown in Figure 3. In general, more female than male subjects reported symptoms. Overall, 54.3% of the individuals reporting symptoms were female. Of individuals reporting symptoms, the largest numbers were seen in the 30–39-year age group. The number of visits remained relatively constant from July 15–19, decreased sharply on Saturday, July 20, and gradually tapered off during the second postspill week (Figure 4). A small increase in the number of visits was noted on Monday, July 29. There were 28 additional visits from August 1–16.

Clinical Manifestations of Exposure. Symptoms most commonly reported at first evaluation are shown in Table I. The symptoms reported at each site were generally similar. The most common symptom reported was headache (63.8%). Nearly one-half of the individuals complained of eye irritation or nausea. Symptoms represented in the "other" category, reported by fewer than 5% of cases, included depression, disorientation, drowsiness, dry mouth, earache, fatigue, fever, hot flashes, irritability, memory reduction, nose bleed, numbness, pain in the arms or legs, ringing in the ears,

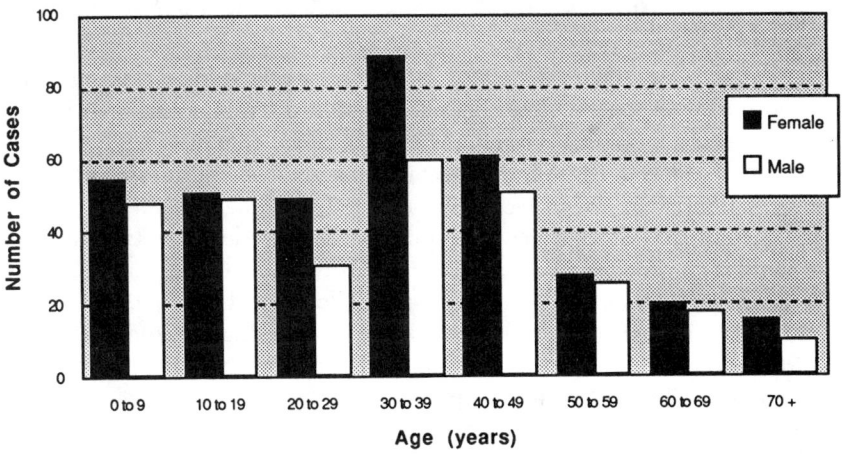

(N = 654, 51 data not given)

Figure 3. Age and gender distribution of cases.

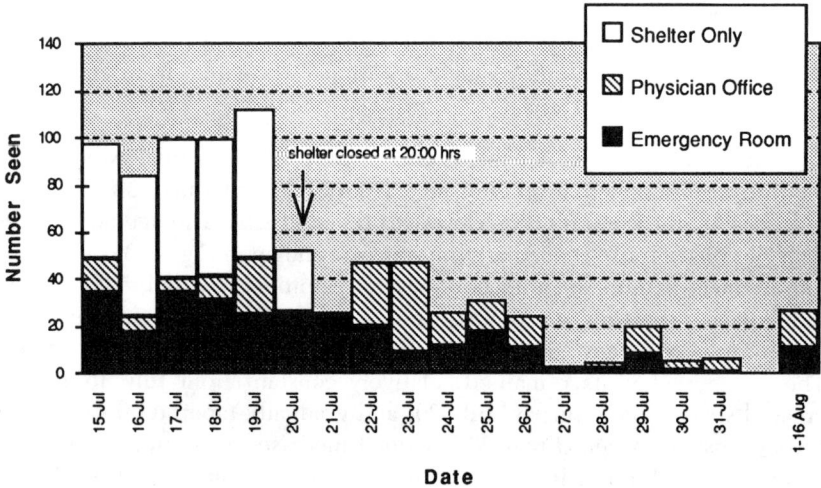

(N = 811, 37 date or site not given.)

Figure 4. Date and site of spill-related medical evaluations.

and sweating. The distribution of symptoms varied little by gender or by age group.

An unexpected finding was the continued reporting of health complaints 1 week after the spill or later. Little difference in the types and distribution of symptoms were reported in those seen within 1 week of the spill, compared with those seen later. However, higher percentages of weakness, diarrhea, cough, rash, and other symptoms were reported among those seen later.

Odors were reported by 247 (35.0%) of cases studies. Most of those reporting an odor (63%) described it as a sulfur or "rotten egg" smell. Other reported odors included insecticide or bug spray, chemical, acid, almonds, ammonia, chlorine, diesel, disinfectant, horseradish, saltiness, and sweetness. A total of 56 cases specifically reported smelling no odors. A group of 402 case reports did not have any information on odor detection and were excluded from analysis. For cases more than 10 years of age, headaches were more frequent among those reporting an odor, but the frequency of weakness, nervousness, and flushing was significantly higher among those reporting no odor. The distribution of other symptoms was generally similar. The mean number of symptoms per case was not influenced by odor detection.

Data on smoking status were limited, but 75 cases were recorded as being current smokers, whereas 48 specifically reported that they did not smoke. These data were missing in 435 cases. The distributions of symptoms

Table I. Number (Percentage) of Cases Reporting Selected Symptoms

Symptom	Number (%) of Cases[a]
Neurologic	
Headache	445 (63.8)
Dizziness	206 (29.6)
Weakness	81 (11.6)
Mucomembranous irritation	
Eye	338 (48.5)
Throat	293 (42.0)
Nasal	161 (23.1)
Gastrointestinal	
Nausea	322 (46.2)
Diarrhea	176 (25.3)
Abdominal pain	131 (18.8)
Vomiting	104 (14.9)
Respiratory	
Shortness of breath	189 (27.1)
Chest tightness	156 (22.4)
Cough	97 (13.9)
Wheeze	81 (11.6)
Dermatologic	
Rash	95 (13.6)
Itching	78 (11.2)
Miscellaneous	
Metallic/odd taste	50 (7.2)
Nervous	33 (4.7)
Flushing	31 (4.4)
Chills	22 (3.2)
Other	264 (37.9)
Mean number of symptoms per case	4.8

[a] $N = 697$.

on the basis of smoking status were similar, although for nearly all symptoms, the frequencies were lower in the nonsmokers. The only symptom in which this difference was statistically significant was shortness of breath. Smokers also tended to report more symptoms per case, but this was not statistically significant.

Hospitalizations. Seven individuals were admitted to the Mt. Shasta Hospital during the first week after the spill, for possible exposure-related illness. Four were respiratory-related admissions, three of whom had preexisting respiratory disease, and two cases of possible syncope (fainting). Cardiac arrhythmia (bigeminy) and disorientation were noted in one individual who probably received a particularly intense exposure through involvement in initial cleanup activities at the spill site. All patients were discharged by July 28. No other known spill-related hospital admissions occurred after this date. No human fatalities were attributed to the spill.

Pregnancies. Eight pregnant women from the spill area were identified during this study. No adverse pregnancy outcomes were recorded. However, two women in their first trimester elected to have a therapeutic abortion, in part because of concerns about exposure effects.

Health Impact of the Spill on Dunsmuir. A total of 498 (70.6%) of the 705 cases listed their city of residence as Dunsmuir. Of these, 391 (78.5%) had home addresses that could be located within or directly adjacent to the Dunsmuir city limits. For the 107 remaining cases, a street address could not be identified, or, in a few isolated cases, the address was much further south or north of the city limits.

Rates of symptom reporting for cases living within the Dunsmuir city limits are shown in Table II. A total of 290 cases, or approximately 13.6% of the Dunsmuir city population, based on 1990 census data, sought medical evaluation for symptoms attributed to the spill during the first month after

Table II. Dunsmuir City Population and Number (Percentage) Reporting Symptoms by Gender and Age

Parameter	Dunsmuir Population[a]	Dunsmuir Cases[b] n (%)
Total	2,129	290 (13.6)
Gender[c]		
Male	1,039	129 (12.4)
Female	1,090	158 (14.5)
Age (Years)[d]		
0–9	306	42 (13.7)
10–19	300	34 (11.3)
20–29	210	41 (19.5)
30–39	289	62 (21.5)
40–49	295	46 (15.6)
50–59	188	22 (11.7)
60–69	255	15 (5.9)
70+	286	13 (4.5)

[a] 1990 census figures.
[b] Only cases in which home located within city limits.
[c] Gender not given for three Dunsmuir cases.
[d] Age not given for 15 Dunsmuir cases.

the incident. Rates of symptom reporting appeared slightly higher among female than male subjects (14.5 versus 12.4%). Larger differences were seen on the basis of age group. Approximately 20% of the 20–29-year and 30–39-year age group population reported symptoms, whereas less than 10% of those in the 60–69-year and over-70-year age groups reported symptoms.

Rates of symptom reporting for consecutive 300-ft zones adjacent to the Sacramento River were calculated (Table III). Total population included in the analysis (2539) is higher than the Dunsmuir city population, because the analysis includes census tracts adjacent to the Dunsmuir city limits. Overall, 391 (15.4%) of the Dunsmuir area population reported symptoms based on 1990 census tract data. Rates were highest in the 0–300-ft zone (21.2%), decreased and remained relatively constant at 12–15% for the 301–600, 601–900, and 901–1200-ft zones, and increased again to 19.8% for the 1201–1500-ft zone. No significant change in distribution occurred when attach rates were stratified on the initial symptom onset date. Rates based on age and gender distribution could not be calculated, because information on these variables within the individual zones was not available.

Exposure Predictions Based on the Fate and Transport Model. As the metam sodium plume traveled south toward Lake Shasta, the pesticide decomposed slowly in the dark and rapidly in sunlight (Figure 5). In the 6 h required for the plume to reach Dunsmuir, only about 5% of the metam sodium decomposed. Between 09:00 and 15:00, when the metam sodium plume was 5–12 mi down the canyon from Dunsmuir and near the communities of Castella, Sweet Briar, Sims, and Gibson, the majority of the pesticide was consumed by photodecomposition. By sundown of the first

Table III. Dunsmuir Area Population and Number (Percentage) Reporting Symptoms by Distance of Home from the Sacramento River and Predicted MITC Concentrations

Distance from river (ft)	Dunsmuir Population[a]	Dunsmuir Cases n (%)	Predicted MITC Concentrations[b]
0–300	406	86 (21.2)	88
301–600	637	87 (13.6)	39
601–900	689	99 (14.4)	26
901–1200	590	76 (12.9)	20
1201–1500	217	43 (19.8)	15
Total	2539	391 (15.4)	

[a]Based on 1990 census block figures, includes areas surrounding city limits.
[b]The MITC concentrations are those predicted for the early morning hours of 7/15 when the metam-sodium plume passed through Dunsmuir. Higher and more uniform exposure concentrations probably occurred later that day in Dunsmuir when the MITC production rate was greater and southerly wind flow developed.

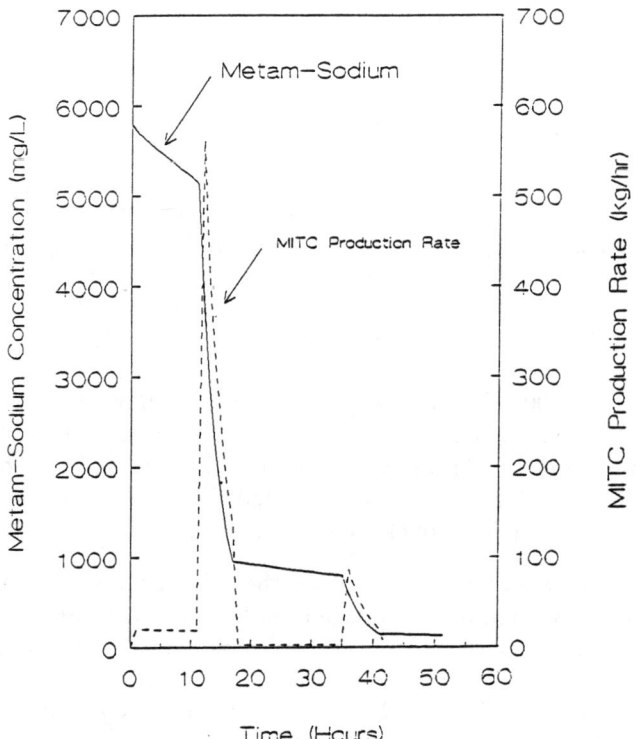

Figure 5. Metam-sodium decomposition and MITC production in the Sacramento River predicted by the environmental fate model.

day, approximately 85% of the compound had photodecomposed, and after the second day of sunlight exposure, metam sodium was almost completely gone. The calculated metam sodium concentration in river water reaching Lakehead and Doney Creek was less than 200 mg/L. This estimate does not take into account dilution due to feeder streams along the 45-mi course of the river, plume dispersion, or decomposition by other abiotic or metabolic processes, all of which would further reduce the concentration.

MITC production was strongly influenced by sunlight. According to the model, immediately after the spill, emissions were about 20 kg MITC/h, but rates as high as 560 kg MITC/h occurred when the sunlight reached maximum intensity. Another MITC pulse is seen the next day, but MITC emissions on July 16 were only about 15% of those on July 15. A total of 2400 kg of MITC volatilized from the river, less than 10% before 9:00 on July 15, >75% during the daylight hours of July 15, and approximately 11% on July 16.

As the plume traveled through Dunsmuir in the predawn hours of July 15, MITC levels were highest nearest the river and dropped off abruptly because of atmospheric dispersion. For example, between 150 and 1350 ft, MITC exposure concentrations dropped from 0.088 to 0.015 ppm. Exposures at this time were strongly influenced by the proximity of Dunsmuir residences to the river. The duration of this exposure corresponds to the time required for the plume to pass, approximately 1 h.

Both the magnitude and duration of Dunsmuir exposures may have been greater later that day when the MITC emissions increased. A southerly, upslope breeze occurring at any time during midday or into the afternoon (confirmed by data from the National Weather Service) would have carried MITC up the canyon to Dunsmuir. These meteorologic conditions are common in mountain valleys: In the evening and early morning, cold-air drainage occurs; in the late morning, as the atmosphere is heated, it becomes unstable and the wind direction reverses. MITC exposure concentrations calculated for Dunsmuir under these conditions depend on the timing of onset of the southerly flow, which also affects the arrival time of airborne MITC in Dunsmuir. During peak sunlight intensity, the metam sodium plume was between 8 and 16 km downstream. The predicted MITC exposure concentrations were as follows: 156 ppb, 10:30 arrival time in Dunsmuir; 89 ppb, 11:45; 59 ppb, 13:00; 35 ppb, 14:10; 21 ppb, 15:20; and 15 ppb, 16:30.

The midday MITC exposure scenario described would have persisted as long as the southerly breeze continued, during which time exposures exceeding MITCs irritant threshold are most likely to have occurred. Another important feature of this scenario is that MITC exposures in Dunsmuir would not have been strongly influenced by location. For example, exposures would not have been stratified over short distances from the river. Moreover, exposures (and symptom complaints) might have been expected for communities upstream from Dunsmuir or even upstream from the spill site (e.g., Mt. Shasta).

The highest MITC exposure concentrations may have occurred 12–18 mi downstream from the spill when sunlight had its greatest impact on metam sodium decomposition. Individuals in the vicinity of the plume at this time (i.e., emergency personnel, workers investigating or cleaning up the spill, press people, bystanders, and residents) may have been subjected to much higher peak exposures than Dunsmuir residents. It is unlikely that MITC exposure concentrations at the river's edge ever reached the MITC odor threshold of 0.100 ppm in Dunsmuir, although other odorous products, including H_2S, were probably frequently detected.

Discussion

In this investigation, self-reported exposure was most commonly associated with nonspecific neurologic complaints (headache and dizziness) and irritation of the eye, respiratory tract, gastrointestinal tract, and skin. Symptoms in nearly all cases were not severe enough to require hospitalization. Of the seven people hospitalized, three had exacerbation of preexisting respiratory disease. Exposure effects on pregnant women are undetermined. Although study found no cases of spontaneous abortions or birth abnormalities, the spill did influence two women to obtain a therapeutic abortion.

Over 15% of those living in the Dunsmuir area reported health effects from the spill. Male and female residents were affected similarly. However, older individuals appeared less likely to report symptoms than younger ones. Explanations for the age differences in symptom reporting could include increased tolerance to irritant effects among older individuals or differences in the degree of exposure among different age groups; alternatively, older individuals may have been less able or willing than younger individuals to seek care for their symptoms.

For the limited number of individuals for whom a smoking history was known, smokers tended to have higher individual symptom rates and a higher number of symptoms overall. This finding may indicate that smokers were more susceptible to irritant effects of exposure or that an unknown interaction exists between smoking and exposure.

Perhaps the most unexpected finding of the investigation was the occurrence of symptom reporting days and even weeks after the spill, when the possibility of exposure to metam sodium breakdown products was unlikely. There were several anecdotal reports of individuals who left the area after the spill, returned when MITC concentrations were not detectable, and then had a recurrence of symptoms. Others reported a recurrence of symptoms following rainstorms two to three months after the spill (citizen testimony at community meeting, November 5, 1991).

Reasons for the later symptom reports are unclear. Our exposure model suggests that significant exposures in excess of the reported MITC irritant threshold only were possible for Dunsmuir residents within the first 24 h after the spill. Symptoms reported 1 week or more after the spill did not appear to differ substantially from those reported initially, except that late symptom reporters had a significantly higher proportion of weakness, diarrhea, cough, and rash.

Several possible explanations exist for the late symptom reports. First, either because of anxiety or because of an interest in formally documenting their symptoms for legal purposes, many of the later medical visits may have represented follow-up visits for initial symptoms that were resolving. Second, residents may have attributed to the incident the development of new health problems, which actually were unrelated to the spill. Third, some of the late symptoms could be attributed to psychological trauma, similar to that seen with posttraumatic stress disorder (23). Certainly, as a result of the spill, area residents were subjected to numerous potential stressors, such as fear of unknown health effects from exposure, sudden evacuation from their home, and helplessness in watching the destruction of a once-thriving river, which may have been their source of economic livelihood. These stressors theoretically could manifest themselves as depression, irritability, difficulty concentrating, sleep disturbances, and fatigue. Fourth, later symptom reports could represent either slowly resolving or chronic problems related to the spill. Finally, previously unrecognized or underestimated toxicologic effects of metam sodium and its breakdown products or sensitization of certain individuals to these chemicals could exist. Low levels of MITC were reported in vegetation samples taken by the CDFG near Dunsmuir as late as August 1991. Although unlikely, reexposure to metam sodium breakdown products weeks after the spill cannot be entirely discounted.

A major limitation of the study is that it is not known with certainty which compounds were responsible for the observed health effects, although toxicologic data indicates that MITC is the most likely candidate. Because reliable air data for the first 2 days after the spill are not available, the model predictions could not be validated with environmental data. In addition, no chemical tests or bioassays can confirm or quantify past MITC exposure in humans.

Because few cases reported the characteristic horseradish odor of MITC, it would appear that MITC levels were below the MITC odor threshold, 100 ppb. Odor reports suggest that hydrogen sulfide was present in a concentration above its odor threshold, 1 ppb. Perhaps the stronger hydrogen sulfide odor masked the MITC odor. Whether hydrogen sulfide was present in concentrations high enough to produce irritative symptoms is not known. However, given that hydrogen sulfide was not expected to be present at higher concentrations than MITC, hydrogen sulfide is a less potent irritant than MITC, and hydrogen sulfide's irritant threshold is almost 10,000 times higher than its odor threshold, it appears that most symptoms were secondary to MITC exposure.

Dunsmuir attack rates did not appear to be strongly stratified with distance from the river. Therefore, consistent with the exposure modeling, exposures exceeding MITC's irritant threshold are most likely to have occurred during the midday upslope air flow. Exposures (and symptom complaints) might be expected for communities upstream from Dunsmuir and even for the spill site itself, as was the case. The highest MITC exposure concentrations may have occurred in sparsely populated areas 12–18 mi downstream from the spill, when sunlight had its greatest impact on metam sodium breakdown. Individuals in the vicinity of the plume at that time may have been subjected to higher peak exposures than Dunsmuir residents.

An interesting prediction from the fate model is that the timing of the spill may have been critical in determining both human health and environmental effects. Had the derailment occurred very early in the morning (i.e., 03:00) the resulting MITC concentrations in Dunsmuir could have been 30-fold higher, because the plume would have reached the town when MITC production was greatest. Had a similar spill occurred in the middle of winter,

MITC exposures to residents along the river would have been considerably lower, although the biological effects to Lake Shasta might have been more severe. A 10 °C drop in water temperature would slow metam sodium's cleavage reaction by 50%, and, more important, in low-intensity winter sunlight, metam sodium's photochemical half-life would increase from 2.5 to 7.4 h—approximately 70% of the metam sodium would have arrived at Lake Shasta.

Conclusions

The Cantara incident had a devastating environmental impact on the affected portion of the Sacramento River. Unfortunately, the full extent of human health effects has not been determined. Results of other follow-up studies are pending. The suddenness and resulting chaos of the event made systematic study difficult. The epidemiological data in this report is observational, uncontrolled, and subject to several types of biases. Information was obtained from different sources. Because of the primary need to treat and triage cases, responding medical personnel could not always obtain or record information for each case in a consistently complete manner.

Although the investigation cannot conclude that cause and effect relationships exist between the spill and reported health problems, several findings suggest causality. Over 700 individuals sought medical attention for symptoms they believed were related to exposure to the spilled chemical, a highly significant finding given the relatively low population of the area. Those who may have been affected but did not or could not seek medical care were not included in this investigation. Therefore, it is likely that this number is an underestimate of the total number of individuals affected. Although some individuals who experienced symptoms were undoubtedly missed by this study, the data presented should provide a reasonable estimate of the range and severity of symptoms experienced. Because the data represent health effects reported as they occurred, the study also is less likely than subsequent retrospective studies to be affected by recall bias.

Symptoms were consistent with exposure to irritant gases, of which MITC appears most likely. Irritation of the eyes, gastrointestinal tract, respiratory tract, and skin occurred. Nonspecific neurologic complaints, such as headache and dizziness, were common. Seven hospitalizations were recorded. No fatalities or adverse pregnancy outcomes were observed. Gender, age, and smoking status appeared to affect symptom reporting.

Some symptoms occurred at chemical concentrations below odor thresholds. In the Dunsmuir area, exposures appeared to occur on either side of the river and at distances of up to ¼ mi from the river. Detailed information on the airborne constituents and their concentrations is not available. Because no fatalities and few hospitalizations occurred, it is unlikely that high-level exposures occurred, as supported by the exposure modeling. Peak exposures occurred within 2 days after the spill; yet for unknown reasons, new symptoms were reported many days and weeks after the spill.

Application of a simple metam sodium environmental fate and transport model suggests that MITC emissions from the spill were very low (e.g., 20 kg MITC/h) as the metam sodium plume passed through Dunsmuir early in the morning of July 15. MITC exposure concentrations were below 0.1 ppm and dropped off abruptly with distance from the river. Sunlight had a major impact on metam sodium decomposition, consuming about 75% of the pesticide during midday of July 15. MITC emissions at midday were as high as 560 kg/h. A southerly (upslope) breeze anytime during the middle of the day or afternoon would have produced MITC exposure concentrations of between 0.156 ppm (metam sodium plume 8 k distant) and 0.015 ppm (plume 16 k away). The midday exposures resulting in this scenario would be relatively uniform in Dunsmuir and could account for symptom complaints upstream from the spill site.

The findings document that metam sodium's volatile decomposition products are irritants in humans and supplement data on occupational exposure to MITC and metam sodium. The symptoms seen in Dunsmuir residents are consistent with exposure to MITC vapors or to other irritant gases. Both the modeling information and the case reports on odor perception suggest that exposure concentrations to MITC were below 200 ppb. This case study demonstrates that in spite of limited resources and by using medical records and questionnaires, valuable human toxicology data can be obtained from epidemiological studies of accidental human exposures.

Acknowledgments

We thank Terri Barber, SCEHD; Richard Sun, Dena Mangiamele, Lynn Goldman, Tim Lomas, Rachel Broadwin, Theresa Saunders, Robert Stephens, Thomas Mischke, CDHS; Dennis Shusterman and Richard Jackson, California Environmental Protection Agency; Diane Wakeham and Mary-Claire Miller, student assistants from the University of California at Berkeley.

References

1. Mahon, D. M., President Southern Pacific Transportation Company. Statement before the U.S. House of Representatives Government Activities and Transportation Subcommittee of the Committee on Government Operation; Oversight Hearing on Train Derailments and Toxic Spills, Washington, DC, October 3, 1991.
2. U.S. Department of Commerce. Economics and Statistics Administration, Bureau of the Census. *1990 Census of Population and Housing. Summary Population and Housing Characteristics of California.* U.S. Government Printing Office: Washington, DC, August 1991.
3. Benghiat, I.; Bowers, J. G.; Lukes, G. E.; Giolito, S. U.S. Patent 2 863 803, 1958.

4. *Questions and Answers about Metham Sodium;* U.S. Environmental Protection Agency: Washington, DC, October 7, 1991.
5. Miller, D. M.; Latimer, R. A. *Can. J. Chem.* **1962**, *40*,, 246–255.
6. Turner, N. J.; Corden, M. E. *Phytopathology* **1963**, *53*, 1388–1394.
7. Joris, S. J.; Aspila, K. I.; Chakrabarti, C. L. *Anal. Chem.* **1970**, *42*, 647–651.
8. "Metham Sodium Aqueous Photolysis at 25 °C." ICI Americas, Inc.; Location, 1990; C. J. Spurgeon, BASF 90/5038, Report RR 90–091B.
9. Worthing, C. R., Ed. *The Pesticide Manual. A World Compendium*, 9th ed.; British Crop Protection Council: Farnham, Surrey, England, 1991; p 557.
10. Ellenhorn, M. J.; Barceloux, D. G. *Medical Toxicology. Diagnosis and Treatment of Human Poisoning.* Elsevier: New York, 1988; p 837.
11. *MSDS Reference for Crop Protection Chemicals*, 3rd ed.; Wiley: New York, 1990; pp 752–753.
12. Nesterova, M. F. *Gig. Sanit* **1969**, *34(5)*, 33–37.
13. Nihon Schering, K. K.; Shinogi and Company, Ltd. *J. Pest. Sci.* **1990**, *15*, 297–304.
14. Mal'tseva, L. M.; Savviatova, N. I.; Chemnyi, A. B. *Gig. Tr. Prof. Zabol.* **1976**, *7*, 53–54.
15. Brook, S. M.; Weiss, M. A.; Bernstein, I. L. *Chest* **1985**, *88*, 376–384.
16. Mullins, F. G. P.; Kirkbright, G. F. *Analyst* **1987**, *112*, 701–703.
17. Dean, A. G.; Dean, J. A.; Burton, A. H.; Dicker, R. C. *Epidemiological Information Version 5;* Centers for Disease Control: Atlanta, GA, 1990.
18. Hwang, S. T. *Land Disposal Toxic Air Emissions Guidelines;* Office of Solid Waste, 1981; Turner, B. D. *Workbook of Atmospheric Dispersion Estimates;* U.S. Department of Health, Education, and Welfare: Washington, DC, revised 1969.
19. Zepp, R. G.; Cline, D. M. *Environ. Sci. Technol.* **1977**, *11*, 359.
20. Zepp, R. G. *Environ. Sci. Technol.* **1978**, *12*, 327.
21. Draper, W. M.; Wakeham, D. E., submitted for publication in *J. Agric. Food Chem.*
22. *ARC/INFO*, Version 5.01; Environmental Research Institute: Redlands, CA, October 1990.
23. *Diagnostic and Statistical Manual of Mental Disorders*, 3rd ed., revised; American Psychiatric Association: Washington, DC, 1987.
24. California Environmental Protection Agency report on the Cantara spill, final draft pending.

RECEIVED for review September 3, 1992. ACCEPTED revised manuscript March 16, 1993.

16

Access to Data for Epidemiological Studies

Ralph R. Cook[1], Sandra L. Tirey[2], Nanette W. Spadacene[3], and Mary A. Woodbury[1]

[1]Epidemiology Unit, Dow Corning Corporation, Midland, MI 48686–0994
[2]Chemical Manufacturers Association, 2501 M Street, N.W., Washington, DC 20036
[3]The Dow Chemical Company, 1803 Building, Midland, MI 48674

> Epidemiological research is dependent on data on exposures, health outcomes, and confounders (alternative causes for the specific diseases of interest). In addition, it often requires a means to tie these three entities together. This is usually achieved by means of personal data, such as name, birthdate, and social security number. In some projects, epidemiologists can gather the data they need directly from the study subjects; but in many projects, they must extract it from records developed for purposes other than health research. Various barriers are associated with access to these secondary sources. Some are related to concerns about privacy and confidentiality. Others involve litigation, economics, logistics, and even philosophy. This chapter explores the problems of data access and proposes some solutions.

EPIDEMIOLOGICAL RESEARCH IS DEPENDENT ON DATA: data on exposures, on health responses, and on confounders (i.e., other causes for the health response of interest). To conduct meaningful work, the database must be relatively complete; that is, there should be little selection bias. In addition, the data must be valid: There should be minimal misclassification of exposures, health responses, or confounders. In particular, because epidemiology is a study of comparisons, there should be minimal differential misclassification; that is, the relative validity of the data should be equivalent for both the exposed and the unexposed, or for both the diseased cases and the healthy controls.

Observational researchers must deal with biases with which most experimentalists have little experience. Humans are not exposed to one agent at a time at some predetermined controlled level, but rather to a multitude of agents simultaneously in an often ill-defined fashion. In trying to understand the potential impact of one agent on human health, the effects of others (confounding variables that confuse the interpretations) must be addressed. Similarly, the pressures that lead to inclusion in any given study by both cases and controls (selection) must be understood and controlled in study design, analysis, or interpretation. Finally, methods of classification or measurement not only must be valid; they must be consistent, or misclassification—particularly differential misclassification—will result. Any of these biases can have a significant impact on the calculated risk estimates and consequently on the interpretations that are derived from them.

In some research, epidemiologists have the luxury of gathering data directly from the study subjects via questionnaires, collection and analysis of blood or urine, or tests of physiological function. They can design the collection procedures to gather a sufficient amount of data and to manage all the phases of data quality.

However, most environmental and occupational epidemiologists must use secondary sources of data that are extracted from records developed by others for reasons unrelated to human health research. For example, the data may come from death certificates, outpatient medical records, personnel files, residential directories, tax records, industrial hygiene reports, sales receipts, or process descriptions. Furthermore, even if the data are valid, occupational and environmental epidemiologists must gain access to it.

Gaining access can be difficult. In some cases, the logistics for obtaining data are extremely cumbersome. In other cases, administrative barriers are based on privacy and confidentiality. Privacy of the individual and confidentiality of data are both extremely important issues in our society. Yet epidemiology, by its very nature, requires invasion of privacy and compromise of confidentiality. Nearly every epidemiological study involves a dynamic conflict that must be resolved.

This paper touches on validity but focuses primarily on access—in particular, the barriers to access. It also summarizes steps being taken in both the public and private sectors to overcome these barriers and thereby to facilitate the more effective and efficient development of epidemiological information.

There are differences among data, information, and knowledge. Knowledge is applied information and information is analyzed data. However, it does not follow that any and all data, even if valid, can be converted to valid information. How data are collected, their focus, and their rigor are all important components that are all too often overlooked.

For example, about 5 years ago in Alberta, Canada, an evaluation of population health statistics identified an alarming and abrupt 50% increase in cancer rates. For about 2 months, this was headline news with various allegations about the risks being associated with life-style factors, or waste dumps, or oil products—all the standard concerns (1). Eventually, it was revealed that the cancer rates had been miscalculated (2).

Cancer rates are simply the number of people (the numerator) in a given population (the denominator) who develop cancer. To make any sense, the numerator must be a subset of the denominator. In this situation, the numerator of the cancer rates, the number of cases, had been obtained from Fort Saskatchewan as well as the surrounding county both before and after the so-called cancer epidemic; however, the denominator data, the number of people in the population under study, had been determined in two different ways. Because of a change in the political boundaries halfway through the study period, in the "before" calculation the number had included those who lived in Fort Saskatchewan and the surrounding county. In the "after" calculation, it had included only those in the county. The "epidemic" was not a result of an increase in cancer. It was simply a result of using data inappropriately, in this case data that were quite valid when analyzed on an element-by-element basis.

This particular mistake obviously triggered and reinforced the public's chemophobia. One has to ask how many other "cancer epidemics" have been a result of similar unintentional, or perhaps even intentional, misuses of valid data?

Obviously, epidemiologists need valid data in order to do meaningful environmental health research. These data often come from different sources, each with its own idiosyncrasies. If investigators are not directly involved with the primary data collection, they must be aware of these differences and adjust for them in study design, analysis, or interpretation. Even if data are known to be valid, epidemiologists must have access to those data for analysis.

Epidemiologists must have access to a lot of data, because epidemiology, as a scientific tool, tends to be a blunt instrument. In aggregate, it can be highly accurate, but it is usually not very precise. Therefore, interpretations of cause and effect in epidemiology depend more on the consistency of results across multiple studies than on the presence or absence of statistical significance or the statistical power of one study. To use the analogy of target shooting, the results of individual studies tend to scatter like pellets of buckshot. What is important is the pattern of results of multiple studies and not necessarily the findings of any single study.

Further complicating the picture is the fact that much epidemiological research is a hypothesis-generating exercise. The data are explored to identify clues that need further investigation. Data are culled to determine the possible. However, more research usually must be done to identify the highly probable. During the initial stages, it may be expedient to identify a plethora of associations to be probed further. During subsequent stages, it is more important to focus, to test hypotheses, and thereby to develop a smaller body of information that either suggests real cause and effect or conversely, if insufficiently compelling evidence of a causal association, suggests that further work should cease and resources should be reallocated to another issue.

Paradoxically, at the same time our society is demanding answers to an ever-increasing array of complex health concerns, it is becoming more difficult to gain access to the data needed to develop the information required to address these concerns. Too many of these barriers are of a regulatory or legal nature.

Study Group

In surveying the basic types of data, the epidemiologist must first decide what group should be studied. This requires answers to the following questions: How should they be identified? What comparison group should be used? What are the sources of data?

In the occupational setting, it is possible to use personnel lists, work histories, department assignments, and the like to develop registries of all the individuals ever employed. To ensure that some block of former employees is not missing from the master registry, it can be cross-checked with the Social Security Administration's 941 files (3). The 941 files provide an annual quarter-by-quarter summary of all those for whom the company has made a contribution. However, these sources give data only on whether and when someone was employed by a particular company. If the company has multiple production lines, they provide no indication of what the employees did at the company or their potentials for exposure. Additional data resources are therefore needed.

In community epidemiology, the problem is even more complex. It has been estimated that at least 4 million people in the United States live within a mile of one of 725 hazardous waste sites on the Environmental Protection Agency's National Priority List (4). In theory, it is possible to develop a registry of who lives where by going to the property tax rolls; but these may list the names of absentee landlords and not include the names of the resident tenants. Alternatively, one could use telephone directories, but not everyone has a telephone. One could go house to house and gather names, but this is a very labor-intensive endeavor and still may not identify everyone. For example, some communities in California contain a number of illegal aliens. These people are reluctant to participate in data-gathering exercises, especially if they must give their names to a government agency.

The issue is further complicated by the fact that the focus of the health research may be a chronic disease such as cancer in which a time lag of years or decades can occur between the exposure and the clinical identification of the cancer. In developing the list of those in a study of chronic disease, the epidemiologist is less interested in who is exposed currently than in who was exposed historically. Unfortunately, the population of the United States is highly mobile. It has been estimated that more than 10% of the population change their zip-codes every year. Finding people can be an exercise in "shoe-leather" epidemiology.

Going back to the occupational setting and given a list of the ever-employed exists, the next problem is determining the sublist of all who were ever exposed. Sometimes it is possible to develop these lists (perhaps even to the extent of defining those with low, medium, and high potential for exposure) by cross-referencing department lists with job descriptions, process information, and industrial hygiene reports. However, this presumes that such information has been gathered and furthermore retained. In the United States, business tends to focus on current survival or future planning. Retention and administration of historical records is expensive. In these times of "downsizing", records that serve no obvious business function are destroyed.

When Eula Bingham was the head of the Occupational Safety and Health Administration (OSHA), this problem was brought to her attention. She mandated that certain records be retained for a period of 30 years after employment (5). This policy seemed to ensure that data would be available to the National Institute for Occupational Safety and Health (NIOSH), the agency charged with doing industrial epidemiological studies. Some think the regulation fell short of the mark. Because it tacitly promoted the destruction of postemployment records older than 30 years, those with shorter terms of employment and therefore shorter durations of exposure were eliminated from long-term chronic disease studies. As a consequence, the epidemiologists' ability to explore the important dimension of dose–response was restricted. In support of epidemiological research, some companies have developed the policy of keeping records for 75 years from date of first employment (Dow Chemical, personal communication). For long-term employees, the OSHA regulation and the Dow Chemical policy have basically the same effect. However, for shorter-term employees, the Dow Chemical policy ensures that more useful data are available.

Exposures

No study of cause and effect can be complete without some implied or explicit estimate of exposures. In the occupational setting, epidemiologists rely heavily on integrating work histories, process records, and industrial hygiene reports. The first two allow the specific determination of who were likely exposed to what (the agents of exposure) and when this was likely to have occurred. The third, the industrial hygiene records, provide an idea of the levels of exposure. Unfortunately, the industrial hygiene records must be used with some caution, because much of the data is collected for compliance purposes. Again, a regulation promulgated for purposes of public health garbles the intent. By law, mandated industrial-hygiene measurements must be taken without regard to the use of personal protective equipment. In other words, two employees standing side by side, one in a self-contained "moon suit" and the other in street clothes, must by law have the same level of exposure recorded. Compliance-oriented industrial hygiene provides measures of potential exposure, not of actual exposure. To obtain a measure of actual exposure, to avoid the problems of interpretation, and to convert industrial-hygiene data to industrial-hygiene information that is useful for epidemiological research, a knowledgeable industrial hygienist must actively participate in the project. Of course, this individual's participation must be done in a fashion that minimizes differential misclassification, but that usually can be handled by a well-designed protocol.

Most industrial-hygiene measurements are of the environment, not the biological entity. They do not measure what enters the worker's body or what enters the likely organ of concern. In the future, data gathered by biomonitoring techniques may prove useful in addressing this shortcoming.

In spite of the limitations, exposure measures in the industrial sector can be extremely useful, but only if they are gathered with epidemiology in mind. In contrast, the exposure measures in community epidemiology can be nebulous. The exposure measures may be modeled in some statistical fashion, or degree of exposure may be determined by distance from the site but without consideration for major variables such as topographical features, wind direction, or hydrology.

In many situations, the exposures are defined by the study participants themselves via a questionnaire. This approach can be fraught with problems. For example, do you remember what you had for lunch three weeks ago; do you remember what chemicals you used in college? On the other hand, if you were told that working with a certain chemical would likely cause a painful death in the future unless you immediately received an antidote, might you not try to recall whether or not you had used the chemical? Selective remembering is referred to as recall bias. Exposure data gathered by questionnaire are particularly subject to it. Furthermore, if the situation is particularly controversial, there may be differential recall bias whereby those with problems do a better job of remembering their past exposures than do those in the comparison group. The result is gross overestimates of risk.

A number of epidemiological studies have attempted to use self-reporting of exposures. Bond et al. (6) investigated this practice and found it to be of questionable validity. For example, the ability to recall one's usual and customary job, the most common job held during the time of employment, varied from 40 to 100%, but the ability to recall specific chemicals used in these jobs went down to single-digit percentages. The opportunities for underestimating risk if nondifferential recall exists or for grossly overestimating risk if differential recall exists are enormous.

Because rigorous exposure data on industrial cohorts are not readily available, especially to researchers outside the private sector, a recent effort has been made to computerize the occupational information on death certificates. An evaluation of the utility of this approach to documenting exposures has yielded most discouraging results (7). At best, this information might be used on a limited basis to generate hypotheses, but it has no utility for testing hypotheses. Furthermore, such exercises lead to two other problems: (1) These types of studies are often given more credence than they deserve; (2) they often develop hypotheses that cannot be tested. Some think that if more rigorous data are not available, this approach should not be used; if more rigorous data are available, this approach may not be needed.

Confounders

In simplified terms, epidemiologists must evaluate the relationship between exposure and health response while simultaneously taking into consideration the impact of confounding variables, the alternative exposures that might cause the same health response. For example, if a study were being done of the potential association between chemical X and lung cancer, it would be necessary to consider the smoking habits of those being studied. If not, any statistical association between chemical X and lung cancer might be confounded, or confused, by smoking.

A confounder has both a biological and a statistical dimension. First, it must be an alternative cause for the health outcome under study; that is, it may be a confounder in any study of that health outcome. Second, it is a confounder only if it is unequally distributed in that particular study. Therefore, whether or not it is a confounder is study-specific. Furthermore, whether the distribution is unequal in a meaningful fashion is often a value judgement. Making those determinations is part of the art of epidemiology. Stated another way, a confounder is just another type of exposure, but one that is confusing or confounding the interpretations of the possible associations of interest. Confounders have all of the problems and limitations that have been enumerated for exposures, and some others.

In the occupational setting, the exposures of interest can be monitored by the industry in some reasonable fashion. However, the confounders can be exposures that take place outside the work environment, such as smoking, drinking, sexual habits, or using illicit drugs. It may be difficult to obtain data on these confounders. In the community setting, any effort to get information on either exposures or confounders can be even more complicated and usually has to be done de nova. Irrespective of whether the epidemiology is performed in the industrial sector or in the community, there must be some forethought about what might be a confounder to ensure gathering the proper data.

Health Outcomes

The third type of data involves health responses. Many occupational medical departments gather clinical health data on active employees via automated multiphasic health testing. In addition, most companies of any size have health insurance for active employees and retirees. A smaller number have set up surveillance programs based on mortality data. Problems related to access and validity can occur, irrespective of source.

For example, the multiphasic clinical data usually are collected only on active employees who choose to participate. Not only does this mean data on former employees are not available, possibly introducing a selection bias, but among those to whom the examinations are available, there may be an additional issue of self-selection—but selection in which direction? Are those with concerns about their health selecting themselves into the process to obtain more information so that they can make reasoned life decisions about diet, exercise, or smoking? Are they avoiding the company-sponsored health examinations because they are concerned that identification of a health problem might cause them to lose their jobs?

Insurance records can provide data on incidence of disease but only for those enrolled in the program, usually some subset of the active employees and retirees. Insurance records do not provide data on employees not enrolled or on those who have left the company. Nor do they provide data of uniform quality. Insurance records are economic instruments. They provide a "trigger" for the payment of health care providers. They are not primarily designed for epidemiological research.

To surmount the limitations of the health insurance records, some industrial epidemiologists have used them as triggers to set up disease-specific registries (8). A claim for cancer, for example, triggers a communication to the health care provider to produce more information. In some situations, it has been possible to develop procedures in conjunction with hospital tumor registries to acquire the sophisticated data. However, in other situations, hospitals have refused to share any information, even when the epidemiologists have had signed releases from the patient (Dow Chemical, personal communication).

We suspect that the refusals are prompted by three things: (1) It is inconvenient for the hospitals to cooperate. In many hospitals, the retention of records is not an orderly process, especially for the files from many years ago. Any effort on the hospital's part to find these records and selectively extract the information requested can be an expensive operation. (2) Among better-organized hospitals, many simply destroy any inactive records that are over 7 years old. (3) Hospitals are concerned about the potential liability associated with sharing data. It is easier for them simply to refuse to cooperate. It would be helpful if procedures were implemented to allow researchers improved access to hospital records, perhaps by explicitly exempting hospitals from liability associated with breach of privacy if they share data with professional epidemiologists performing health research.

In selected sites, it has been possible to negotiate collaborative research with state tumor registries (9). This has allowed access to sophisticated tumor-incidence data on both active and former employees, or at least on those still living in the catchment area of the registry. It would be useful if truly national tumor registries were developed and were available to epidemiologists, perhaps with access procedures based on acceptable protocols.

Much of the work of industrial epidemiologists has involved mortality as the outcome of interest. They have used both the Social Security Administration (SSA) and the National Death Index (NDI) to determine who is alive and who is dead. Unfortunately, the SSA has lately chosen to reduce severely their epidemiological research support activities (10). In years past, it was possible to obtain information from the SSA that would place former employees in three basic groups: living, dead, and lost to follow-up. Resources could then be allocated to finding the death certificates for the deceased and to tracking down those lost to follow-up in order to determine their vital status. The SSA now provides only two basic categories: dead and lost to follow-up (which includes all the living and truly lost to follow-up). As a consequence, more resources are needed to determine the vital status, and a major inefficiency has been introduced into the system.

In theory, the NDI, a function within the National Center for Health Statistics, is specifically designed to support epidemiological research. Although the service this government agency provides is superb, unfortunately it is somewhat constrained in both its time frame and charter, which constrains its use for chronic-disease research (11). At present, the NDI has data from as early as 1979. For data before that time, it is necessary to query the SSA records for clues as to who is dead and the state in which the death occurred.

Even when the SSA was providing more complete disclosure, the data were incomplete. Information on the state in which the last financial transaction took place was given. However, the state of last financial transaction is not necessarily the state in which the death certificate is stored. After a person dies in North Dakota, the next of kin could process the forms in Michigan. The SSA records would suggest the death certificate should be obtained from Michigan. However, any query placed there would obviously fail, and other search strategies would have to be used. For those employed in the northern states, the next step often would be to query a number of other states, for example, the "snowbird" states like Florida and Texas, states to which retirees travel during the winter. Tracking down the records can be a time-consuming process. In this example, however, even that strategy would fail, and a laborious state-by-state search would be necessary. To help streamline this process, it would be helpful if information could be made available, perhaps through the NDI, to all epidemiologists, perhaps indirectly from the Internal Revenue Service records, as it is to NIOSH. It also would be helpful if the NDI could change its current limit of 1979 to at least another 10 years earlier. In this age of cancer phobia and interest in other diseases of long latency, 10–15 years is too short a time span.

Of course, whether the investigators use data from the SSA or the NDI, they still have to go back to the states for copies of the death certificates. Some states, such as Texas and Michigan, are a joy to work with. Others have been quite the opposite.

For example, to obtain a death certificate from the state of South Carolina it has been necessary to sign a "warranty of confidentiality", which said among other things that the primary investigator and his or her heirs were fully liable for any damage caused by improper use of the data, whether such improper use resulted from their own actions or from the actions of any person or institution obtaining access to the data (State of South Carolina, personal communication).

In South Carolina, an industrial epidemiologist was not allowed to sign as an agent for the company. The researcher was to be personally responsible. If a court required sharing the death certificate with the plaintiff's lawyers and they misused the data, the epidemiologist could be held financially responsible. In fact, a liberal interpretation would suggest that if the state of South Carolina misused the data after the death certificate had been provided to an outside researcher, the epidemiologist could be held financially responsible. Needless to say, this release form posed a significant barrier to access.

In occupational and environmental epidemiology, it is often important to do the research quickly in order to place into perspective a highly emotional situation. Various companies have set up registries designed to allow quality studies to be done quickly. However, some states have policies that specifically prevent that from happening. For example, Pennsylvania requires a separate application and written approval each and every time a study uses Pennsylvania data (State of Pennsylvania, personal communication), even if new data are not being requested. Although this policy is reasonable, it is still one more impediment to be addressed before the research can proceed.

In some situations, it is necessary to obtain approval from an Institutional Review Board (IRB) before the information on the death certificate can be used to contact the attending physician or the next of kin to get additional information. The problem is that the IRB must be approved by the National Institutes of Health (NIH). This seems to be a reasonable requirement, but it contains a "catch-22". An industry-based IRB can conform to all the recommendations of the NIH but still be unable to obtain NIH approval because the institute will certify only research organizations that receive NIH funds. If a research group is self-sufficient and does not want to complete all the paperwork necessary to obtain such funds, it cannot receive NIH approval.

It has been argued that these procedures are needed for social reasons, that is, to protect the privacy of the individual and that of the next of kin. The paradox is that in most states, death certificates are public documents, available to anyone who walks in off the street, as long as the documents are requested one at a time. Ordering more than one for health research triggers a series of time-consuming procedures, which often do not appear in any regulations or written department policies.

For example, in a situation involving access to copies of nine death certificates, the state of New Jersey required a protocol. The requirement of a protocol is reasonable, because among other things, it ensures that procedures will be in place to protect against any undue invasion of privacy. However, the New Jersey bureau specifically required that the protocol has to be reviewed and approved by a union. In this particular situation, eight of the nine people had been salaried employees, and the ninth had retired before the current union became certified. Thus, none of the nine were "represented" by the union.

We would like to see the whole process streamlined. For example, it would be more efficient if access to death certificates for research purposes was under the control of NDI. Perhaps this could be accomplished, even to the extent of providing the death certificate and the assigned nosology codes, the unique numerical codes for each specific cause of death as defined in the International Classification of Diseases. To financially support the data-gathering effort, some fee should be sent back to the originating states, but this could be part of the NDI's charges for the service.

Discussion and Conclusion

A number of regulations and related government agencies have been developed with the intent of improving the information that is available for reasoned decisions about human health and the environment. In theory these regulations are good, but in practice they often are subverted. The Toxic Substances Control Act *(12)* is an example. It contains what some consider "gotcha" legislative requirements. The legislation does not require industry to do epidemiological research. However, if research is conducted, the legislation effectively forces an overinterpretation of the data under some severe time constraints.

Furthemore, the regulations or their interpretations change over time *(13)*. Industry epidemiologists recently have been required to go back and report, under the pressure of new government interpretations, findings that had been published in the open scientific literature more than 10 years previously with each report involving a fine of up to $15,000 and a cap of $1 million. Under such pressures, it should be no surprise that many companies are reluctant to set up epidemiological programs.

One other impediment to epidemiological research is the litigation process and the role plaintiff lawyers play in it. It is the impression of many in the scientific community that when a situation becomes controversial, plaintiff lawyers are close at hand. They disrupt the research process in one of two ways. First, if the attorneys debrief their clients before they are contacted by the epidemiologists, they exacerbate the recall bias, which leads to overestimates of risk. This has been labeled as the "litigation bias" *(14)*. Second, some also have the impression that plaintiff attorneys frequently find it easier to argue their case on the basis of allegation, innuendo, and vague hypotheses rather than facts, at least facts as scientists understand them *(15)*. At times, it appears that lawyers advise their clients not to cooperate in any study that might develop information prejudicial to their case (that would support the lack of cause and effect). Within the ethics of the legal profession, this may be a reasonable position. Epidemiologists find it offensive.

It is also troublesome because decisions seem to be made about complex scientific issues by a process that does not seem to place great emphasis on scientific analysis. At a recent legal symposium, one attorney even argued that science has no place in a court of law *(16)*. The social implications of this position are enormous. For example, some pharmaceutical companies have refused to market efficacious drugs, because the potential cost of defense against litigation, often specious litigation, is greater than the anticipated profit of the product. To help introduce more objective science into the legal process, the Chemical Manufacturers Association (CMA) has supported course work at the National Judicial College designed to teach judges about the proper role of science in the courtroom.

It may sound as if we are arguing that it is impossible to conduct occupational and environmental health research. That is not our intent. In most cases, it is possible to find ways to overcome these obstacles, but they can be very time-consuming and resource-draining.

Many health researchers think it is important to make better information available more rapidly so that better decisions can be made within a reasonable amount of time. The Agency for Toxic Substances and Disease Registries has been set up to address this issue on a community basis *(17)*. Gathering the necessary data and converting these data to useful information constitute an awesome responsibility that ATSDR researchers will need help to accomplish.

In the private sector, a number of companies have set up in-house epidemiological research units. However, obviously many more units are needed to build the body of available valid data to a size sufficient to address the important questions from the perspective of the consistency of study results. For this reason the CMA established a pilot project called the Epidemiology Resource and Information Center (ERIC) *(18)*. This group is developing and publishing a series of monographs designed to make it easier for individual companies to develop in-house expertise.

One of these monographs is the *Occupational Epidemiology Resource Manual (19)*. It provides an overview for the nonepidemiologist manager, and some practical tips for professionals interested in beginning or expanding a program in industry. At least one graduate school, the University of Massachusetts at Amherst, is using it as a resource manual. It should also prove useful to consultants and government research organizations. Another monograph is the *Guidelines for Good Epidemiology Practices for Occupational and Environmental Epidemiologic Research*. These Good Epidemiology Practices (GEPs) are to epidemiology what the Good Laboratory Practices, the GLPs, are to toxicology. Both of these CMA documents are available to anyone who is interested, and the GEPs have been published *(20)*.

Most of the epidemiologists who work full time in industry, at least those in the petroleum and chemical industries, belong to a group called the Industrial Epidemiology Forum (IEF). This group also has published a series of monographs. The most recent, developed in conjunction with Tom Beauchamp (a bioethicist at the Georgetown University Institute of Ethics), has been incorporated into the proceedings of an IEF conference called "Ethics in Epidemiology", which was published as a supplement to the *Journal of Clinical Epidemiology (21)*. It specifies in some detail the obligations epidemiologists have to subjects, society, funders and employers, and colleagues.

In summary, a number of problems confront occupational and environmental epidemiologists as they strive to develop meaningful information about the potential health effects associated with exposures to toxic chemicals. Some of these problems are related to issues of validity, and many more involve data access. Paradoxically, society is demanding more information while, albeit sometimes indirectly, preventing ready access to the very data needed to develop this information. Epidemiologists have found ways of accomplishing their work in spite of the system; however, it would be helpful if the system were modified to make access to valid data more time- and resource-efficient.

References

1. Hicks, G.; Saloway, T. *The Edmonton Sun* **1987**, *April 15*.
2. Roberts, W. *The Record* **1987**, *June 10*.
3. Marsh, G. M.; Enterline, P. E. *J. Occup. Med.* **1979**, *21*, 665–670.
4. *Principal Findings about the Linkages between Releases of Hazardous Substances and the Impact on Public Health: Public Health Impact;* Biennial Report

Volume II, October 1986–December 1988; Agency for Toxic Substances and Disease Registry (ATSDR): Atlanta, GA, 1990; pp 57–59.
5. *Fed. Regist.* **1980**, *45*, 35212–35228.
6. Bond, G. G.; Bodner, K. M.; Olsen, G. W.; Burchfiel, C. M.; Cook, R. R. *Appl. Occup. Environ. Hyg.* **1991**, *6*, 521–527.
7. Olsen, G. W.; Brondum J.; Bodner, K. M.; Kravat, B. A.; Mandel, J. S.; Mandel, J. H.; Bond, G. G. *Am. J. Ind. Med.* **1990**, *17*, 465–481.
8. Pell, S.; O'Berg, M.T.; Karrh, B. W. *J. Occup. Med.* **1978**, *20*, 725–740.
9. Bond, G. G.; Austin, D. F.; Gondek, M. R.; Chiang, M.; Cook, R. R. *J. Occup. Med.* **1988**, *30*, 443–448.
10. Dupree, E. A. *Epidemiol. Monitor* **1990**, *11*, 10–12.
11. Bernier, R. H.; Mason V. M. *Epidemiol. Monitor* **1991**, *12*, 400.
12. *Fed. Regist.* **1978**, *43*, 11110–11148.
13. *Fed. Regist.* **1991**, *56*, 19514–19536.
14. Woodside, F. C.; Lydon, D. R. *N. Engl. J. Med.* **1983**, *308*, 1604.
15. Huber, P. W. *Sci. Am.* **1992**, *266*, 132.
16. Dunleavy, D. M. Presented at the Conference on Breast Implant Litigation, New York, June 1, 1992.
17. Baller, J.; Carlo, G. L.; Sund, K. G. *Bur. Nat. Affairs Environ. Rep.* **1991**, *3*, 1951–1956.
18. Bernier, R. *Epidemiol. Monitor* **1989**, *10*, 1–3.
19. Pastides, H.; Mundt, K. A. *Occupational Epidemiology Resource Manual*; Chemical Manufacturers Association: Washington, DC, 1991.
20. Chemical Manufacturers Association *J. Occup. Med.* **1991**, *33*, 1221–1229.
21. Beauchamp, T. L.; Cook, R. R.; Fayerweather, W. E.; Raabe, G. K.; Thar, W. E.; Cowles, S. R.; Spivey, G. H. *J. Clin. Epidemiol.* **1991**, *44* (Suppl. 1), 151S–169S.

RECEIVED for review September 3, 1992. ACCEPTED revised manuscript January 27, 1993.

17

The Successes and Failures of Environmental Epidemiology

Raymond Richard Neutra

Environmental Health Investigations Branch, California Department of Health Services, 5900 Hollis Street, Suite E, Emeryville, CA 94608

> *Environmental epidemiologists and analytical chemists increasingly find themselves in a sort of intercultural marriage. They make an odd couple but an effective one. A review is presented on how the instincts and training of the two disciplines differ and why so much of the practice of environmental epidemiology may seem counterintuitive to a chemist. The successes of environmental epidemiology can come from discovering a hazard, its effective dose, or the environmental conditions that deliver it to susceptible populations. Another kind of success is showing that a particular dose and route of exposure do not seem to explain a disease outbreak. Examples of both kinds of successes are given, and the conditions that promise each are discussed. It is argued that despite the overreaction of the media to individual studies, epidemiology rarely leads to false-positive results in the regulatory process. However, epidemiology often fails to identify hazards, particularly if they convey less than a 1.5-fold relative risk.*

CHEMISTRY IS THE QUINTESSENTIAL EXPERIMENTAL DISCIPLINE, whereas epidemiology only *observes* the natural world and can control nothing. Chemists are reductionists who try to understand phenomena in the simplest terms in systems that are isolated from external influences. Epidemiologists watch the world as it is and accept the interplay of numerous factors influencing phenomena they are trying to study *(1)*. Chemists can see subtle changes with concentrations at the part per trillion level, whereas epidemiologists can see only the influence of factors that cause a 1.5-fold or higher change in the diseases they study. Chemists, because they are experimentalists, are perfectionists and attempt to control all external factors to the greatest possible degree. Epidemiologists have no hope of controlling anything; they simply must learn to outsmart nature and to find clever ways to negotiate the inherent messiness of the problems with which they are dealing. Chemists make the greatest effort to measure everything that is relevant to their experiments. Epidemiologists assume that many unmeasured factors are at work and are satisfied with simply bracketing the likely magnitude of these unseen effects. Chemists have their own theory and method, whereas epidemiologists rely on the theory of biology, use the statistical method, and apply the methods to a variety of disease problems as they arise.

It is no wonder that the chemist asks the epidemiologist, "How can you possibly use such an inexact nonexperimental science that has so little unique theory to drive environmental policy, which is targeted to helpful chemicals that I synthesize or use?" This chapter discusses the successes and failures of environmental epidemiology, beginning with the successes.

What do epidemiologists mean by success? One kind of success is a "true-positive" finding. In this instance, a human health risk is associated with a certain dose range of a chemical received by way of an existing route of environmental exposure; epidemiologists, with the help of supporting scientists, (1) detect that hazard, (2) identify the dose that is hazardous, and (3) find the exposure route by which that kind of dose actually reaches people. An example of this would be the events in California of July 4, 1985. Reports of three individuals who developed vomiting and diarrhea within half an hour of eating a watermelon led to a flurry of telephone calls to emergency rooms around the state and to the realization that the pesticide aldicarb had been illegally used at a ranch growing watermelons. This agent had traveled up the root of the watermelon into the fruit and achieved a dose sufficient to produce this illness. The hazard was identified, the dose level was bracketed with the help of analytical chemists, and, after serious detective work, the route of exposure was determined. Early detection of additional cases, by means of calls to emergency rooms around the state, led to public warnings and to the ban of watermelon sales, which curtailed what would have been a widespread epidemic of poisonings (2).

Another kind of success would be the demonstration of a "true-negative" result. In this case, no actual human health risk from a chemical in a particular dosage and exposure route existed. Epidemiologists and supporting scientists, after careful study, do not find any excess risk from that exposure and that dose. Epidemiologists do not provide generic exoneration of an agent, because they always study the agent in a particular real-life situation, which may not be relevant to another situation. An example of a true-negative result would be our study of an episode of well-water contamination with trichlorethane (TCA) (3). We showed that a cluster of miscarriages and birth defects reported in a neighborhood near a leaking underground tank could not be caused by that leak, because another neighborhood with more contamination had not experienced an increased risk and because within the "cluster" neighborhood miscarriages and birth defects had not occurred at the homes using the most contaminated water (4). Thus, TCA at close to 1 ppm in water was not associated with miscarriages or birth defects. This does not mean that TCA at higher doses would not pose a hazard.

Epidemiology has worked successfully under the following four conditions:

1. When a sudden chemical exposure in which an illness is produced within a very short time occurs, such as in the clearly documented tragedy at Bhopal in India, where deaths and severe illness resulted from a chemical explosion (5). Another example is the watermelon episode discussed earlier (2).

2. When under typical conditions and in very short periods of time, a researcher can correlate a change in a physiological function to the change in an environmental pollutant. An example would be the study of asthmatic school children at summer camps in areas where it can be documented that as the air pollution from ozone fluctuates, bronchoconstriction also fluctuates (6).

3. When an association has been made between long-term exposures and long-term functional effects. The most famous example is the demonstration that exposure to fairly low levels of lead, as judged from the accumulation of lead in the teeth, is associated with poor performance on standard psychological tests (7).

4. When occupational studies have linked relatively high levels of chemical exposure to the incidences of cancer in workers. Their results have been extrapolated downward to the lower exposures in the general environment in order to estimate the potential cancer risks from environmental exposures. Usually, it is assumed that a linear dose relationship exists (8).

For extrapolation, we assume that the rate ratio (i.e., the ratio of the rate of disease in exposed and unexposed individuals) or the rate difference (i.e., the difference between rates in exposed and unexposed individuals) increases linearly with doses. At the higher dose levels, we have actual observations and are interested in knowing how that curve would progress if the dose were to become less and less and approach zero. Complicated statistical models exist to do this, but they involve a "leap of faith", drawing either a straight line or some curvilinear function from the zone of actual observations down to zero. Sometimes this has to be done on the basis of human evidence because no animal model for carcinogenesis is available;

this has been the case for arsenic and asbestos. In other situations, we have both animal and human data. In the case of cadmium (9), it seems that humans are less sensitive than animals.

What about failure? Do studies ever give a false-positive result, "crying wolf" when none is really lurking in the vicinity? Interestingly, there appears to be no generally accepted example of a clearly unnecessary regulation being instituted on the basis of faulty epidemiological studies.

Why is it difficult to find a clear-cut failure of the false-positive variety? There are a number of reasons. Scientists are trained to avoid false-positive results. This is never clearly stated, but epidemiologists, as scientists, must recognize that their credibility within society will not survive if they cry wolf too often. This is reflected in the statistics epidemiologists use. The statistics grew out of some ad hoc decisions by Ronald Fisher (10) in the 1920s and are only now beginning to receive critical reevaluation. The conventional tests of statistical significance set their false-positive and false-negative errors (the alpha and beta errors) in such a way as to avoid the false-positive ones. Nonetheless, we must acknowledge countervailing practices that could lead to false-positive results. The first is the examination of many hypotheses that all have a chance to be statistically significant. The second is the tendency to publish positive results. However, the regulatory process itself provides ample opportunities for interested parties to challenge the evidence underlying any regulation, and industry has the resources to hire scientists to subject any epidemiological study to the severest criticism. Regulation requires more than one study to show a strong result. So regulation itself is unlikely to produce a false-positive finding. The same is not true for the media. The media do not refrain from seizing the flimsiest epidemiological study to attract their readers and from making a story of it.

The opposite failure is the error of the false-negative result, the "wolf in sheep's clothing". This is an error that epidemiology, by its very nature, must make constantly, because the diseases studied are rare, because a background rate is influenced by many other factors, and because effects that may still be of social consequences are small. Thus, by studying a group that is too small, we fail to see the effect of interest. Economic constraints may also prevent us from following an exposed group for a sufficient period of time, particularly for diseases like cancer, which has an incubation period that may be as long as 40 years. Epidemiologists also make the error of a false-negative finding because of their reluctance to study controversial topics that defy conventional knowledge. Current examples are electromagnetic fields and their potential health effects and the controversial assertion that some individuals suffer multiple chemical sensitivities to very low levels of chemicals. If we do not study these problems, we can never resolve them one way or the other.

Epidemiologists may fail to detect an existing problem when they examine a complex mixture in which some unidentified component may be hazardous. One classic example is the problem of the "sick building syndrome" or indoor air pollution. Evidence exists that workers in buildings that use artificial ventilation are more likely to have episodes of ill-defined and bothersome symptoms among workers; yet every conceivable chemical and microorganism in the ventilation system has been measured and no offending culprit has been identified. We seem to be looking here for a needle in a haystack and cannot find it. The great temptation is to label the workers with symptoms as simply having a psychosomatic problem, because we have not found a measurable causal factor in their building. Epidemiologists can ascertain that something is happening, but they need their colleagues in the laboratory to help find that needle.

Finally, just because an epidemiologist cannot implicate a cause, does not mean it is unimportant. Epidemiologists tend to assume that an environmental problem does not exist because an epidemiological study has failed to demonstrate it. A good example of this was the controversy surrounding saccharin use in the late 1970s. Some preliminary epidemiological studies (11) had shown an association between bladder cancer and the use of saccharin in soft drinks. An animal study had shown that animals fed high doses of saccharin developed bladder cancer. Finally, an enormous case-control study was conducted nationwide (12) and did not show an association between saccharin and bladder cancer. Some epidemiologists interpreted this finding to mean that rats develop bladder cancer from saccharin but humans do not. In fact, by using the animal data to extrapolate to the human doses, a relative risk of only 1.01 would have been expected on the basis of the usual saccharin dose in diet drinks. The study itself, although enormous, only had the ability to detect a relative risk of 1.1 and could not have detected an effect as small as 1.01. Now, it might be said that if it could only cause a relative risk of 1.01, that is, a 1% excess in the incidence of bladder cancer, and that this effect would be negligible; in fact, that 1% would represent about 800 cases of bladder cancer per year in the United States. You can imagine what trial lawyers would have done if they could have proven that a specific diet soda was actually producing 800 cases of cancer a year. So, although this is an excess of social interest, it is an excess that epidemiologists simply will never be able to see, even if it did exist.

Summary

This chapter reviewed some of the differences between chemistry and epidemiology and the successes and failures of environmental epidemiology. Epidemiology has been most successful when focused on large acute health effects and has often been aided when chemists have provided ways to assess levels of harmful chemicals in the environment or in human tissues. It is less successful when an agent causes a less than 50% increase of disease incidence and when the interval between the time of exposure and the onset of disease is long. Epidemiology is a useful instrument that has definite limitations. The art is in using this instrument appropriately. People do not look down on the sledge hammer—they just don't use it to elicit knee-jerk reflexes.

References

1. Kleinbaum, D. G.; Kupper, K. L.; Morgenstein H. *Epidemiological Research;* Van Norstrand Reinhold: New York, 1982.
2. Goldman, L. R.; Smith, D. F.; Neutra, R. R.; et al. *Arch. Environ. Health* **1990**, *45*, 229–236.
3. Wrensch, M.; Swan, S. H.; Lipscomb, J.; Epstein, D.; Fenster, L.; Claxton, K.; Murphy, P. J.; Neutra, R. R. *Am. J. Epidemiol.* **1990**, *131*, 283–300.
4. Wrensch, M.; Swan, S. H.; Murphy, P. J.; Lipscomb, J.; Claxton, K.; Epstein, D.; Neutra, R. R. *Arch. Environ. Health* **1990**; *45*, 210–216.
5. Andersson, N.; Muir, K. M.; Salmon, A. G.; Wells, C. J.; Brown, R. J.; Prunell, C. J.; Mittal, P. C.; Mehra, V. *Lancet* **1985**, *i*, 751–762.
6. Lippman, M. *J. Air Pollut. Control Assoc.* **1989**, *39*, 672–695.
7. Mushak, P.; Davis, J. M.; Crocetti, A.; et al. *Environ. Res.* **1989**, *50*, 11–36.
8. Hertz-Picciotto, I.; Gravitz, N.; Neutra, R. R. *Risk Anal.* **1988**; *8*, 205–214.
9. Collins, J.; Brown, J. P.; Painter, P. R.; Jamall, I. S.; Zeise, L. A.; Alexeeff, G. V.; Wade, M.J.; Siegel, D. M.; Wong, J. J. *Regulat. Toxicol. Pharmacol.* **1992**, *16*, 57–72.
10. Fisher, R. A. *The Design of Experiments;* London, Oliver and Boyd; Edinburgh, United Kingdom, 1935.
11. Office of Technology Assessment. "Cancer Testing Technology and Saccharin"; Office of Technology Assessment: Washington, DC, October 1977.
12. Hoover, R. N.; Strasser, P. *Lancet* **1980**, 837–840.

RECEIVED for review September 3, 1992. ACCEPTED revised manuscript February 11, 1993.

INDEXES

Author Index

Auletta, Angela E., 89
Bond, James A., 137
Cimino, Michael C., 89
Cook, Ralph R., 231
Craigmill, Arthur L., 53
Dong, Michael H., 189
Draper, William M., 53, 189, 209
Freeman, Caroline S., 175
Grossman, Elizabeth A., 175
Hewitt, David J., 209
Houk, Vernon N., 1
Kershaw, William C., 39
Klaassen, Curtis D., 39
Kreutzer, Richard A., 209
Maldonado, George, 29
McClellan, Roger O., 137
Needham, Larry L., 121
Neutra, Raymond Richard, 245

Papanek, Paul J., Jr., 189
Recio, Leslie, 137
Ross, John H., 189
Shane, Barbara S., 65
Sowers, MaryFran, 21
Spadacene, Nanette W., 231
Spivey, Gary H., 9
Stephens, Robert D., 189
Tamburro, Carlo H., 153
Tirey, Sandra L., 231
Vine, Marilyn F., 105
Wetzlich, Scott, 53
Woloshin, Kimberley A., 189
Wong, John L., 153
Woodbury, Mary A., 231
Yuan, Bo, 153
Zhang, Peide, 153

Affiliation Index

California Department of Health Services, 53, 189, 209, 245
California Department of Pesticide Regulation, 189
Chemical Industry Institute of Toxicology, 137
Chemical Manufacturers Association, 231
The Dow Chemical Company, 231
Dow Corning Corporation, 231
Los Angeles County Department of Health Services, 189
Louisiana State University, 65
Occupational Safety and Health Administration, 175

Procter and Gamble Company, 39
U.S. Department of Health and Human Services, 1, 121
U.S. Environmental Protection Agency, 89
University of California, Davis, 53
University of Kansas, 39
University of Louisville, 153
University of Michigan, 21
University of Minnesota, 29
University of North Carolina, 105
Unocal Corporation, 9

Subject Index

A

Abortifacients, human, 79–80
Absorption
 definition, 44
 dose estimation for risk assessment, 58
 influence of physicochemical properties, 44
 through skin, environmental exposure to synthetic chemicals, 5
Acceptable daily intake, chemicals that exhibit toxicity thresholds, 60

Acid metabolites
 accumulation in tissue groups, 197, 199
 excretion rates, 199
Acquired immunodeficiency syndrome (AIDS),
 relationsip between characteristic and
 disease, 12
Acrylonitrile
 exposure levels, comparison with estimates
 of job rank orders, 161f
 lack of correlation with liver cancer,
 171–172
 molecular epidemiology, 153–173
 symptoms of acute toxicity, 154
 work and exposure history, 157t
Actuarial data, toxicological vs.
 epidemiological studies, 63
Adducts, measurement in humans, 124
Adverse effect level, lead, 6
Aerial application, malathion, 191
Age, loss of kidney function, 184–185
Agency for Toxic Substances and Disease
 Registry
 epidemiological research, 242
 urinary arsenic levels, 117
Agent Orange
 components, 126
 exposure indices, 127–128
 pharmacokinetics in humans, 126–127
Alberta, Canada, miscalculation of cancer
 rates, 232–233
Alcohols, teratogenesis, 78, 81
Aldicarb, in watermelons, 246
Allergic reactions, characteristics, 42
Aluminum, teratogenesis, 80–81
Amaranth, teratogenesis, 74
American Association for the Advancement of
 Science (AAAS), teratogenesis of
 2,4,5-T, 81
American Conference of Governmental
 Industrial Hygienists
 biological exposure indices, 125
 recommendations for carcinogens, 61
 threshold limit value, 60
Ames mutagenicity test, bacterial
 mutagenesis assay, 56
Analytical monitoring for acrylonitrile,
 history of project, 160
Analytical observational studies, types,
 23–26
Anesthetic gases, spontaneous abortion, 79
Angiosarcoma, vinyl chloride, 2
Animal studies
 basis for linking human health risks and
 environmental factors, 2

carcinogens, chronic feeding studies, 3
high doses of toxic agents, 47
selection of species, 47–48
toxicity testing, 48, 56
Aplastic anemia, benzene, 2
Arsenic
 changes in lymphocytes and sperm, 82
 spontaneous abortion, 79
 urinary levels as marker, 117
Arthritis, exposure to chemicals, 2
Artificial ventilation, failure to detect
 source of problem, 248–249
Asbestosis, exposure to chemical or physical
 agents, 2
Aspartame, teratogenesis, 74
Assumptions, quantitative risk assessment, 3
Atropine, side effects, 41
Attributable risk
 definition and illustration, 15
 vs. relative risk, 16t
Availability, marker selection, 110

B

Background levels
 determination of attributable risk, 15
 toxicant found in general population,
 124–125
Benzene, aplastic anemia and myelogenous
 leukemia, 2
Bias
 case-control studies, 26, 35–36
 definition, 17
 epidemiological studies, 30–37
 exposure indices, 123
 litigation bias, 242
 recall bias, 236
 selection bias, 238
 sources, 231–232
Biliary tract, excreted chemicals, 46
Bills of mortality, first death
 certificates, 10
Bioassay, lifetime carcinogenicity studies,
 50–51
Biochemical markers, environmental
 epidemiology, 105–120
Biological exposure indices, American
 Conference of Governmental Industrial
 Hygienists, 125
Biological markers
 collaborative research, 115–117
 criteria for selection, 110–113
 definition, 105, 106, 118

INDEX

dose absorbed, 205
environmental epidemiology, 105–120
measurement in biological samples, 190
relationship to exposure and disease, 106f
types, 106–108
Biological monitoring, evidence of exposure, 58–59
Biological response, markers, 107
Biological variability, severity of the disease process, 13
Biologically based models, toxicity at low doses, 61
Biologically effective dose
assessment of dose–response relationship, 124
individual exposure, 108
marker, 107
Biomonitoring techniques, industrial-hygiene measurements, 236
Biotransformation reactions, toxic chemicals, 45
Birth defects, cluster occurrence, 246–247
Blastocyst, damage from toxic compounds, 66–67
Blood
direct measurement of toxicant or metabolite, 124–126
lead levels, threshold and adverse effects, 6
variability in lipid content, 125
Bloomington, Indiana, PCB contamination, 130
Butadiene
cytochrome P450-dependent metabolism, 142
human health risks, 137–152
metabolism rates, 151
targets for carcinogenicity, 140
Butadiene monoepoxide (BMO), enzyme-mediated conjugation, 143

C

Cadmium
characteristics and uses, 175
cumulative airborne exposure, 182–185
dose vs. probability of kidney dysfunction, 185f
effect on human body, 176–177
environmental sources and exposure, 176–179
kidney dysfunction from occupational exposure, 175–187
time-weighted exposure, 180–181
Caisson disease, exposure to chemical or physical agents, 2
California
aldicarb in watermelons, 246
Medfly eradication campaign, 189–208
metam sodium spill, 209–230
California Department of Fish and Game, fish kills in Sacramento River, 211
California Department of Health Services, epidemiological questionnaire, 215
California Environmental Protection Agency, physiological and biochemical values for malathion, 195
Cancer
1,3-butadiene, 137–152
dioxin, 7–8
exposure to chemical or physical agents, 2
Cancer rates, determination, 233
Cantara metam sodium spill, epidemiological assessment, 209–230
Carbon tetrachloride, immediate toxicity, 41
Carcinogenicity
bioassays, maximum tolerated dose, 62
1,3-butadiene, 137–152
chronic and lifetime studies, 50–51
chronic feeding studies, 3
dioxins and furans, 6–8
expense of studies, 50–51
extrapolation along a theoretical dose–response curve, 60–61
interaction of chemicals or their metabolites with DNA, 138
model based on experience with radiation, 3–4
rodent studies, 49
Cardiac glycosides, persistence in the body, 46
β-Carotene, study of effect on betel quid chewers, 108–109
Case-control studies
definition, 79, 121–122
description and evaluation, 25–26
Causality
associations between risk factors and disease, 9
criteria to judge evidence, 18
difficulty in epidemiology, 17–18
models, 17f
temporal sequence, 25
Centers for Disease Control and Prevention
investigation of clusters, 22
study of cancer mortality, 7
Chemical-bioassay approach, new chimeric plasmid, 163
Chemical classification scheme, Gene-Tox program, 94t

Chemical exposure, cause of illness, 2–3
Chemical Industry Institute of Toxicology, DNA-reactive chemical research strategy, 138
Chemical Manufacturers Association
Epidemiology Resource and Information Center (ERIC) monographs, 243
science in the courtroom, 242
Chemical toxicokinetics, exposure levels, 62
Chemically induced mutagenicity, data evaluation, 89–104
Chemistry, contrasted with epidemiology, 245–246
Chickens, use as test species, 48
Chloracne
halogenated aromatic hydrocarbons, 2
high-level human exposure to dioxin, 7
soil and serum levels of dioxin, 128
Chromatin Ha-*ras* DNA, treatment with acrylonitrile epoxide, 170–171
Chronic carcinogenicity studies, design and objectives, 50–51
Chronic obstructive lung disease, exposure to chemicals, 2
Classification, source of error, 32
Clusters, descriptions, 22
Cohort studies
defining incidence and investigating potential causes of disease, 25
definition, 79, 122
description and evaluation, 23–25
prospective and retrospective, 23–24
selection of individuals to participate, 24
time frame, 24f
Collaborative research, biological markers, 115–117
Comparison estimates, fundamental assumption, 32–33
Computer database, Gene-Tox program, 92
Computer software, SAR testing, 56
Confounders
control, 34–35
epidemiological data, 32–35, 237–238
judging magnitude and direction, 35
markers, 110
Congenital malformations, exposure to chemicals, 2
Conjugation reactions
BMO with glutathione, 148, 149t
excretion into bile, 46
metabolism, 45
Consensus, quantitative risk assessment, 3

Consonance with existing knowledge, causality, 18
Contaminant, pathway from emission to resulting health effect, 122f
Control of experimental conditions, study design, 29
Correlational studies, description and evaluation, 23
Coumarin, central nervous system malformations, 76
Cross-sectional studies
definition, 122
description and evaluation, 22
Cytogenetic effects of butadiene, rats vs. mice, 150

D

Data, sources and access for research, 232
Data-analysis techniques to eliminate confounding, 35
DDT
concentration in fish, 131
human internal dose, 131
sequestration, 44–45
Death certificates
historical reason for development, 10
occupational information, 237
Decompression illness, exposure to chemical or physical agents, 2
Delayed vs. immediate toxicity, 41
Department of Defense, teratogenesis of 2,4,5-T, 81
Dermal exposure
disposition in tissue compartments, 193–194
equations typically used in a PB-PK model, 196–197
physiological and biochemical values for malathion, 195
physiologically based pharmacokinetic model, 194f
Descriptive studies, types, 22–23
Descriptive toxicology, definition, 40
Design methods, occupational and environmental epidemiology, 21–27
Detoxification enzymic reactions, BMO, 150
Developmental toxicity
assays to study mechanism, 70
hydra embryos, 73
ranking chemicals within a family, 71
ratio between toxicity in adult and fetus, 72

INDEX

Devonshire, endemial colic, 11
Diabetes, exposure to chemicals, 2
Dibenzofurans, teratogenicity, 81
Dibromochloropropane, development of sperm, 82
Diethylstilbestrol
  abnormalities in male reproductive tract, 80
  adenocarcinoma, 80
  delayed toxicity, 41
Dioxin
  Agent Orange, 127–128
  cancer, 7–8
  chloracne, 7
  epidemiological studies, 6–8
  need for environmental cleanup, 8
  occupational study, 131–132
  pharmacokinetics in humans, 126–127
  residential exposure, 128
  serum levels in Vietnam veterans, 129t
Disease
  causation mechanisms, 109
  exposure to chemical or physical agents, 2
  markers, 107
  measurement, 12–14
  spectrum, 12f
Distribution, physicochemical properties, 44–45
DNA, chemical carcinogens, 138
DNA adduct, biologically effective dose marker, 107
DNA-reactive chemicals
  human health risks, 137–152
  research strategy components, 139f
Dogs
  toxicity tests, 48–49
  use as test species, 48
Dose estimation
  cadmium inhalation, 180
  malathion, 189–208
  techniques, 58–59
Dose level, route of administration, 49
Dose makes the poison, principle of toxicology, 3
Dose quantification, epidemiological studies, 5
Dose reaching embryo or fetus, 69
Dose–response relationship
  acute and chronic exposure situations, 114
  allergic reactions, 42
  assumptions, 43
  dioxin, 7–8
  mutation, 151

radiation and tumor development, 4
threshold, 56–57
toxicity testing, 39
Dose simulation, assumed time lapse, 204
Dow Chemical Company, occupational record maintenance, 235
Drugs
  human teratogens, 81–82
  side effects, 41
Dunsmuir, California, metam sodium spill, 209–230
Dust, source of lead, 6

E

Ecological fallacy, causal association at the individual level, 23
Ecological studies, description and evaluation, 23
Electric fields, leukemia, 25
Elimination, toxic chemicals, 45–46
Ellis cohort, kidney dysfunction among male workers at a cadmium smelter, 180–181, 184–185
Embryo, period of greatest damage from toxic compounds, 66–67
Embryolethality, reproductive toxicity assays, 69–78
Embryotoxicity, reproductive toxicity assays, 69–78
Emotional problems, exposure to chemicals, 2
Endemial colic, Devonshire, 11
Endonuclease S_1 analysis
  chromatin Ha-*ras* DNA, 171f
  epoxide-modified pSV2neo-Ha-*ras* plasmid, 170f
Enterohepatic circulation, definition, 46
Environmental cohort studies, examples, 23
Environmental contamination, adverse effects on late-maturing functions, 68
Environmental epidemiology
  for chemists, 39–51
  successes and failures, 245–250
Environmental exposure, cadmium, 177–179
Environmental monitoring, assessment of human exposure, 132–133
Environmental Mutagen Information Center, Gene-Tox program, 90
Environmental Protection Agency (EPA)
  Gene-Tox program, 89–104
  negative impact and remedial activities, 43

reference dose (RfD), 60
species selection, 48
Environmental risk, assessment, 1–8
Environmentally induced disease, "black box" approach, 105
Enzymes
  metabolism of butadiene, 141
  teratogenic compounds, 70
Epidemics, methods of field observation, 10
Epidemiological studies
  access to data, 231–244
  basis for linking human health risks and environmental factors, 2
  conflicting results, 4
  health of a population, not individuals, 5
  human carcinogens, 4
  inconclusive results, 5
  interpretation, 29–38
  methods, 9–19
  reproductive loss in humans, 78–83
  role in risk assessment, 1–8
  types, 22t
Epidemiology
  contrasted with chemistry, 245–246
  definition, 9, 105
  shift from infectious to chronic diseases, 10
  vs. toxicology, 53–54
Epidemiology Resource and Information Center (ERIC), monographs, 243
Epoxide modification
  chromatin DNA, 158
  Ha-*ras* plasmid DNA, 157
1,2-Epoxybut-3-ene [butadiene monoepoxide (BMO)], enzyme-mediated conjugation, 143
Estimators of risk, cohort and case-control studies, 27f
Ethanol
  hazard, 40
  teratogenesis, 81
Ethics
  epidemiological research publication, 243
  risk assessment studies, 63
Ethylene oxide, spontaneous abortion, 79–80
Etretinate, craniofacial and limb defects, 82
Exposure
  and health effects pathway, 122f
  effect on reproductive outcomes, 79
  implied or explicit estimates, 235–237
  multiple sources, 16
  need for validation, 3
  self-reporting, 236–237
  timing and duration, 78–79

Exposure assessment
  choice of methods, 132
  fundamental requirement, 57–59
  malathion dose, 204
  markers, 109–110, 117
  measurement, 16–17
  theoretical models, 59
Exposure classification, 57
Exposure concentration, measurement, 59
Exposure index
  dioxin, 131–132
  estimated, 123
  sources of error, 123–124
  test of validity, 127–128
Exposure levels
  chemical toxicokinetics, 62
  dose–response relationship, 114
Exposure to chemicals, causal relation to illness in humans, 5
Exposure to environmental contaminants, measurement of internal dose, 121–135
External dose, total concentration of toxicant in environment, 123–124
Extraneous risk factors, confounding, 32–35
Extrapolation
  animal studies to humans, 82–83
  quantitative risk assessment, 3
  rate ratio or rate difference, 247
  uncertainty in risk characterization, 62

F

Falck cohort, workers at plant producing refrigeration compressors with silver-brazed copper fittings, 181–185
False-positive and false-negative results, 248
Fate and transport model, methyl isothiocyanate concentration predictions, 223–225
Fertility assessment by continuous breeding, test protocol, 77
Fetal alcohol syndrome, abnormalities, 81
Fetal development, physiology, 66–68
Fetotoxic compounds, evaluation, 65–66
Fish
  tests for reproductive toxicants, 74
  uptake of lipophilic methylmercury, 80
Fluoride and bone fracture in Iowa, special-exposure cohort, 24–25

INDEX

Food and Drug Administration (FDA)
  risk-to-benefit paradigm, 43
  species selection, 48
Food chain
  cadmium sources, 176
  effects of cadmium contmination, 178–179
Fort Saskatchewan, cancer epidemic, 233
Frequency of disease, example, 13
Frog embryo teratogenesis assay (FETAX), 74–75
Fumigant, metam sodium, 211
Fungicides, teratogenesis, 78

G

Gasoline, source of lead, 6
Gastrointestinal tract, absorption site, 44
Gene-Tox program, 89–104
Genetic control, hypersusceptibility to succinylcholine, 42
Genetic mutations, ionizing radiation, 3
Genetic toxicology, Gene-Tox program, 89–104
Genotype, variations, 83
Gestation index, reproductive toxicity tests, 76–77
Glomerular filtration, favorable factors, 45–46
Glomerular proteinuria, cadmium-induced renal disease, 177
Glucuronidation, conjugation reaction, 45
Gold standard method, disease diagnosis, 114
Ground troops in Vietnam, exposure to Agent Orange, 127–128
Growth, in utero exposure to lead, 6

H

Ha-*ras* DNA, chemical modification by epoxides, 167–170
Ha-*ras* oncogene
  activation, molecular model, 172
  carcinogenic potential of acrylonitrile epoxide, 153–173
Halogenated aromatic hydrocarbons, chloracne, 2
Hazard
  analytical process of identification, 55
  definition, 40
  evaluation goals, 47
  vs. toxicity, 40
  See also Risk assessment

Hazardous waste site, biological markers, 117
Health care providers, adverse reproductive outcomes, 79–80
Health outcomes, data sources, 238–241
Healthy worker effect
  example of selection bias, 26
  generation of systematic error, 26*f*
Hearing, in utero exposure to lead, 6
Heart disease
  exposure to chemicals, 2
  smoking, 16
Hepatic angiosarcomas, occupational exposure to vinyl chloride, 155
Hepatotoxicants, reversible toxic effects, 42
Hepatotoxicity
  correlation with cumulative acrylonitrile exposure, 162
  worker cohort, 157
Herbicides containing dioxin, pharmacokinetics in humans, 126–127
Histogenesis, functional activity of fetal organs, 67–68
Hospitals, epidemiological research, 238–239
Human biomonitoring, evidence of exposure, 58–59
Human carcinogenesis
  epidemiological studies, 4
  implication, 3
Human data, hazard identification, 55
Human developmental stages, susceptibility to reproductive toxicants, 67*f*
Human embryonic palatal mesenchyme cells, growth inhibition, 72
Human exposure assessment, internal dose, 124–126
Human health, strategy for assessment of risks, 1–8, 137–152
Human neuroblastoma cells, inhibition or stimulation of differentiation, 72–73
Hydantoin, teratogenesis, 75
Hydra, teratogenicity test system, 73
Hydrolysis of butadiene monoepoxide, kinetic constants, 148*t*

I

Idiosyncratic reactions, definition, 42
Illnesses, exposure to chemical or physical agents, 2
Immediate vs. delayed toxicity, 41

Immunological disorders, exposure to chemicals, 2
In vitro assays
  advantages and disadvantages, 76
  reproductive toxicants, 65–87
  short-term, 70–76
  toxicity, 56
In vivo assays
  mammalian studies, 76–77
  reproductive toxicants, 65–87
  toxicity to whole animals, 56
Incidence, definition, 9, 13
Incubation period
  definition, 11
  time-line illustration, 11f
Indoor air pollution, failure to detect source of problem, 248–249
Industrial Epidemiology Forum, monographs, 243
Industrial hygiene records, levels of exposure, 235–236
Infertility, exposure to chemicals, 2
Information bias, epidemiological studies, 30–32
Ingestion, synthetic chemicals, 5
Inhalation, synthetic chemicals, 5
Insecticides, teratogenesis, 78
Institutional Review Board, information on death certificate, 240–241
Insurance records, epidemiological research, 238
Integrated dose, marker selection, 112
Intermediate end points
  biological markers, 117–118
  markers of dose–response relationship, 114
Internal dose
  contaminant, 123
  exposure measurement, 121–135
  marker, 107–108
Interpreting epidemiological studies, basic strategy, 30
Intervention studies
  description, 26–27
  participant compliance with treatment regimens, 108–109
Invasiveness of technique to obtain specimens, marker selection, 110–111
Invertebrate embryos, teratogenicity test system, 73–74
Ionizing radiation, genetic mutations, 3
IQ level
  in utero exposure to lead, 6
  PCBs and dibenzofurans in cooking oil, 81
Irreversible vs. reversible toxic effects, 41–42
Itai-itai, cadmium contamination, 178

J

Japan
  Jinzu River basin, cadmium contamination, 177–179
  Minamata Bay, methylmercury in the food chain, 80

K

Kidney
  chemical elimination, 45
  dysfunction risk from exposure to cadmium, 175–187

L

Laboratory data, hazard identification, 55
Laboratory research, compared to epidemiological research, 115–117
$lacZ^-$ mutant frequency
  butadiene-exposed animals and controls, 146t
  determination, 145–146
Late symptom reports, methyl isothiocyanate exposure, 226–227
Latency
  chronic diseases, 11
  definition, 9
  time-line illustration, 11f
Lead
  adverse effect level, 6
  animal and human studies, 8
  development of sperm, 82
  endemial colic, 11
  epidemiological studies, 6
  in utero exposure, 6
  measurement in blood, 124
  sequestration, 44–45
  spontaneous abortion, 79
Leukemia, exposure to electric and magnetic fields, 25
Lifetime carcinogenicity studies
  design and objectives, 50–51
  in humans, 4

INDEX

Limb bud, teratogenesis, 73
Linearized multistage model, mathematical risk characterization, 61
Lipophilic compounds, measurement of internal dose in blood, 125
Lithium, teratogenic effects, 80–81
Litigation bias, epidemiological research, 242
Liver
 detoxification of butadiene monoepoxide, 150
 tests of workers exposed to acrylonitrile, 164f–165f
 xenobiotic metabolism, 45
Liver angiosarcoma, vinyl chloride, 4
Liver cancer, lack of correlation with acrylonitrile, 171–172
Liver microsomes, preparation for metabolism studies, 142
Local effects
 definition, 41
 vs. systemic effects, toxic reactions, 41
Logistic regression model, dose and dysfunction, 182–185
Los Angeles County, California, exposure to electric and magnetic fields, 25
Lung cancer, impact of cigarette smoking, 15

M

Magnetic fields, leukemia, 25
Malathion
 dermal vs. oral exposure, 199, 201
 doses in California's Medfly eradication campaign, 189–208
 metabolic pathway in humans, 190
Malathion metabolites
 ratio in human urine, 193
 subjects potentially exposed to Medfly eradication sprays, 192t
Mammalian embryos, teratogenecity test systems, 75–76
Mapping
 chemical modification sites based on the use of S_1 endonuclease, 168
 modification of sites in Ha-ras DNA, 155–159, 169f
Markers
 environmental epidemiology, 105–120
 measurement in biological samples, 190
 validation, 113–115
 See also Biological markers, Molecular markers

Mass-balance theory, sum of chemical in tissue compartments, 205
Maternal metabolism, protection of fetus, 69
Maximum tolerated dose, carcinogenicity studies, 50, 62
Measurement error
 information bias, 31
 source, 32
Mechanistic toxicology, definition, 40
Medfly eradication campaign
 health risks of aerial applications, 190
 malathion doses, 189–208
 study materials and methods, 191–197
Memphis, Tennessee, PCB contamination, 130
Mercury, teratogenesis, 80
Mesothelioma, exposure to chemical or physical agents, 2
Metabolism, toxic chemicals, 45
Metabolizing enzymes, teratogenic compounds, 70
Metam sodium
 chemistry and properties, 211–212
 decomposition pathways, 212f
 epidemiological assessment of spill, 209–230
 exposure predictions based on fate and transport model, 223–225
 toxicology, 213
Methodological problems, risk assessment, 61–62
Methotrexate, teratogenesis, 75
Methyl isothiocyanate
 acute health effects of exposure, 209–230
 estimating exposures in Dunsmuir, 217–218
 exposure predictions based on fate and transport model, 223–225
 reported health effects, 225–227
 toxicology, 213
Methyl mustard oil, See Methyl isothiocyanate
Mice
 tumor development with butadiene, 137–152
 use as test species, 48
Microsomal metabolism
 butadiene, 146–147
 butadiene monoepoxide, 147–148
Minamata Bay, Japan, methylmercury in the food chain, 80
Mining, cadmium contamination, 178
Miscarriages, cluster occurrence, 246–247
Misclassification
 biased risk estimates, 57
 hypothetical example, 31–32

using environmental data and
  questionnaires, 3
Models
  exposure estimation, 59
  extrapolation along a theoretical
    dose–response curve, 60–61
  logistic regression, 182–185
  physiologically based pharmacokinetic,
    189–208
Molecular markers
  definition, 105
  environmental epidemiology, 105–120
Monkeys, use as test species, 48
Monte Carlo technique, PB-PK models, 203–204
Mouse embryo limb bud cell culture assay,
  organ culture, 73
Mouse ovarian tumor cells, attachment, 72
Multimedia exposure assessment, 59
Multistage linearized model, quantitative
  risk assessment, limitations, 5–7
Mutagenicity
  experimental study design, 143–144
  extrapolation along a theoretical
    dose–response curve, 60–61
  Gene-Tox program, 89–104
  usefulness of assays, 96
Mutant frequency, butadiene-exposed animals
  and controls, 146–147
Mutation
  induction, 140–141, 149
  ionizing radiation, 3
  role of damaged DNA, 137–152
Myelogenous leukemia, benzene, 2

N

National Academy of Science
  dose–response relationship, 57
  paradigm of chemical risk assessment, 54
National Center for Health Statistics,
  epidemiological research, 239
National Death Index, epidemiological
  research, 239–240
National Health and Nutrition Examination
  Survey, DDT serum levels, 131
National Institute for Occupational Safety
  and Health
  cancer mortality study, 7
  occupational dioxin study, 131–132
National Institutes of Health, information
  on death certificate, 240–241

National Library of Medicine, TOXNET
  database, 90
National Toxicology Program
  analysis of database, 95
  comparison to Gene-Tox program database,
    95–96
  reproductive toxicity test, 77
Nested case-control study, example, 26
Neuromuscular disorders, exposure to
  chemicals, 2
New Bedford, Massachusetts, PCB
  contamination, 130
New Jersey, protocol for death certificates,
  241
Nongenotoxic chemicals, prediction of
  carcinogenesis, 4

O

Oak Ridge National Laboratory, Gene-Tox
  program, 90
Occupational exposure
  adverse reproductive outcomes, 79–80
  cadmium, 175–187
  determining study group, 234–235
  dioxin, 131–132
  estimates, 235–237
  prospective medical surveillance, 153–173
  vinyl monomers as carcinogens, 154
Occupational Safety and Health
  Administration (OSHA)
  acrylonitrile work level, 160–162
  assessment of risk, 186
  logistic regression model, 182–185
  Occupational Exposure to Cadmium: Proposed
    Rule, 180–182
  occupational record maintenance, 235
Occupational surveillance system, basic
  components, 154
Odds ratio
  definition, 25
  hypothetical example, 31
Office of Pollution Prevention and Toxics,
  Gene-Tox program, 89–104
Oncogenes, carcinogenic potential of
  acrylonitrile epoxide, 153–173
Operation Ranch Hand study, dioxin herbicide
  effects in humans, 126–128
Oral administration, toxicity studies, 49
Organ culture, mouse embryo limb bud cell
  culture assay, 73

INDEX

Organ systems, susceptibility to reproductive toxicants, 67f
Organophosphorus compounds, test species, 48
"Ouch-ouch" disease, cadmium contamination, 178
Oxidation of butadiene to butadiene monoepoxide, kinetic constants, 147t

P

Paint chips, source of lead, 6
Paoli, Pennsylvania, PCB contamination, 129–130
Paraquat, accumulation in the lung, 44
Pathology, geographical and historical, 10
Pennsylvania, application for use of data, 240
Persistence, marker selection, 111
Person, definition in epidemiology, 10
Person–place–time method, environmental epidemiology, 11
Pesticide
  metam sodium spill, 209–230
  spontaneous abortion, 80
  urban application, 189–208
Pesticide illness reports, metam sodium spill, 215–216
Pharmaceutical companies, implications of litigation, 242
Pharmacokinetic studies
  bodily processes, 44–46
  definition, 43
Physiologically based pharmacokinetic (PB-PK) models
  malathion dose estimation, 189–208
  Monte Carlo technique, 203–204
  numerical procedure, 194–195
  physiological and biochemical values for dermal exposure to malathion, 198t
  prediction of tissue dose in animals and humans, 193–197
  reduced uncertainty, 62
  simulation of amounts of malathion in tissues, 200t
  time-dependent tissue concentrations, 206
Place, definition, 10
Planaria, effect of toxicants on reproduction and teratology, 74
Plasma cholinesterase, idiosyncratic reaction, 42
Polychlorinated biphenyls (PCB)
  age as predictor of internal dose, 131
  contamination of soil and surfaces, 129–130
  teratogenicity, 81
Postimplantation assay, reproductive toxicity, 75
Preclinical disease, markers, 109
Predictability, short-term tests, 71t
Predictivity, in vitro screen, 70–71
Pregnancy
  detection in early stages, 68
  exposure to chemical substances, 68
  physiology, 66–68
Preimplantation assay, reproductive toxicity, 75
Prevalence
  definition, 9, 13–14
  description and evaluation of studies, 22
  relationship to incidence, 14
Privacy of individual, health research, 241
Proteinuria
  Jinzu River basin, 178–179
  stages, 177
Proximal tubule secretion, plasma, 46
Psychological disorders, exposure to chemicals, 2
Puffer fish, hazard, 40

Q

Quality-control issues, study results, 115
Quantitative risk assessment
  extrapolation, 3
  nonthreshold toxicants, 60–61

R

Rabbits, use as test species, 48
Radiation-induced mutations, tumor development, 3
Radiation sickness, exposure to chemical or physical agents, 2
Railroad transportation accident, metam sodium spill, 209–230
Random allocation, study design, 29
Rank-order estimates
  environmental exposure to acrylonitrile, 156–157
  retrospective validation, 159–162
Rank-ordered index, system for estimating chemical exposure, 154

ras oncogenes, central role in human cancer, 163
Rats
  subacute testing, 49
  tumor development with butadiene, 137–152
  use as test species, 48
Recall bias
  case-control studies, 26
  exposure data, 236
Receptor-mediated activity, dioxin, 6
Reference dose (RfD), Environmental Protection Agency, 60
Regeneration, sensitivity to teratogenic agents, 74
Regiospecific modification of Ha-*ras* oncogene, indicator of health risk, 153–173
Regulations based on overestimates, consequences, 1–2
Regulatory toxicology, definition, 40
Relative risk
  definition, 14–15
  exposed and nonexposed populations, 3
  odds ratio, 25
  vs. attributable risk, 16t
Relative teratogenic index, ratio between toxicity in adult and fetus, 72
Renal excretion, extent of ionization, 46
Reproductive failure, toxic compounds, 68–69
Reproductive toxicity
  hydra embryos, 73
  in vitro and in vivo assays for screening, 65–87
  preimplantation and postimplantation assay, 75
  single-generation tests, 77
  three-generation tests, 76–77
Residential exposure
  dioxin, 128
  metam sodium, 215–229
  toxic waste sites, 129
Reversible vs. irreversible toxic effects, 41–42
Risk assessment
  cumulative cadmium dose, 182–185
  estimates and confidence limits, 54
  kidney dysfunction from exposure to cadmium, 175–187
  principal components, 53
  synthetic toxic substances, 1–8
  toxicology, 53–64
  uncertainties, 61–62, 186
  See also Hazard

Risk characterization, definition, 59
Risk factors, definition, 12, 21
Rodents
  acute toxicity tests, 48
  carcinogenicity studies, 4, 49, 95t
  teratogenecity test systems, 75–76
  use as test species, 48
Route of administration, dose level, 49

S

Saccharin, failure to demonstrate problem, 249
Sacramento River, metam sodium spill, 209–230
Safety factors, threshold toxicants, 59–60
Safety studies, purpose for toxicologists, 39
Salmonella
  bacterial mutagenesis assay, 56
  Gene-Tox database, 92–96
Scarlet fever, defeat of Athens by Sparta, 9
Science policy, quantitative risk assessment, 3
Scientific data, public health benefits, 5
Screening, reproductive toxicants, 65–87
Scrotal cancer, soot, 4
Selection bias
  case-control studies, 26
  epidemiological studies, 35–37
  evaluation and control, 37
  multiphasic clinical data, 238
Sensitivity, in vitro screen, 70–71
Sensitization reactions, characteristics, 42
Sequestration, lead and DDT, 44–45
Serum dioxin levels, chloracne, 128
Seveso, Italy, dioxin exposure, 7, 128, 129t
Sick building syndrome, failure to detect source of problem, 248–249
Simulation packages, physiologically based pharmacokinetic models, 195
Skin–blood partition coefficient, malathion dose estimation, 204
Smoking
  cadmium exposure, 184–185
  heart disease, 16
  lung cancer, 15–16
  teratogenesis, 81
Social Security Administration, epidemiological research, 239–240
Sodium methyldithiocarbamate, *See* Metam sodium

INDEX

Soil, source of lead, 6
Solvents, teratogenesis, 78
Somites, index of teratogenicity, 75
Soot, scrotal cancer, 4
South Carolina, responsibility for use of data, 240
South Wales, correlation between central nervous system malformations and aluminum, 80–81
Special-exposure cohort, rare exposures, 24–25
Species differences, genotoxic and carcinogenic effects of butadiene, 149–150
Species selection, toxicity testing, 48
Specificity
  causality, 18
  in vitro screen, 70–71
  marker selection, 110
Spectrum of disease, definition, 12
Spermatozoa, chemicals affecting normal production, 82
Spontaneous abortion
  chemicals, 79–80
  epidemiological studies, 78–80
  rate determination, 68
Stability in storage, marker selection, 113
State tumor registries, epidemiological research, 239
Storage depots, chemicals in the body, 44–45
Streptomycin, biliary excretion, 46
Stroke, exposure to chemicals, 2
Structure–activity relationship (SAR) analysis 55–56
Study design
  elimination of confounding, 34
  occupational and environmental epidemiology, 21–27
Study group, data source, 234–235
Success, definition in epidemiology, 246
Succinylcholine, idiosyncratic reaction, 42
Suitability of assay, marker selection, 113
Summary exposure statistics, epidemiological studies, 16–17
Supercoiled plasmid DNA, inverted repeats of poly(dG)–poly(dG) sequences, 168
Susceptibility, markers, 107–109
Sweden, rate of miscarriage among petrochemical plant workers, 80
Synthetic toxic substances, health risk, 1–8
Systematic errors, case-control studies, 26
Systemic circulation, entrance pathways, 44

Systemic effects
  toxicity studies, 42–43
  vs. local effects, toxic reactions, 41
Systemically active toxicants, undesired effects, 41

T

Taiwan, PCBs and dibenzofurans in cooking oil, 81
Target tissues, definition, 106
Temporal relationship between exposure and disease, cohort study, 23
Teratogenesis
  chemicals, 80–82
  definition, 68
  evaluation of compounds, 65–66
  reproductive toxicity assays, 69–78
  toxic compounds, 68–69
Tetracyclines, biliary excretion, 46
Tetraethyl lead, local and systemic effects, 41
Tetrodotoxin, hazard, 40
Thalidomide, variations in genotype, 83
Threshold dose, embryo or fetus, 69
Threshold limit value, American Conference of Governmental Industrial Hygienists, 60
Threshold toxic effects, definition, 54
Threshold toxicants, safety factors, 59–60
Time, definition in epidemiology, 10–11
Time sequence, causality, 18
Time to appearance, marker selection, 111
Time-weighted exposure levels
  acrylonitrile, 160
  cadmium, kidney function, 180–183t
Tissue–blood partition coefficient, malathion dose estimation, 204
Tissue culture, assays for teratogens, 72–73
Toxic substances
  fetal development, 66
  health risk, 1–8
Toxic Substances Control Act, epidemiological research, 241
Toxic waste sites, residential exposure, 129
Toxicity
  parameters, 49–50
  reproductive, 65–87
  vs. hazard, 40
Toxicity studies, definition, 43
Toxicity testing
  animal studies, 47–51
  examples, 54
  principles, 47

risk assessment, 57t
tiered approach, 55
Toxicology
  definition, 39
  dose–response relationship, 57t
  major fields of inquiry, 40
  use in epidemiology, 39
  vs. epidemiology, 53–54
Toxicology Data Network System (TOXNET) database, 90, 96–97
Transfection–transformation assay, transforming potential of *ras* DNA, 163
Transforming genes, *ras* oncogene family, 163
Triana, Alabama, DDT residues, 131
Trichlorethane, well-water contamination, 246–247
2,4,5-Trichlorophenoxyacetic acid (2,4,5-T), teratogenesis, 81
True-positive and true-negative results, 246
Tubular cells, transport of organic acids and bases, 46
Tubular proteinuria
  cadmium-induced renal disease, 177
  cumulative exposure to cadmium, 186f
Tumor(s), chronic feeding studies of laboratory animals, 3
Tumor-incidence data, epidemiological research, 239
Tumor information, development of model, 138
Tumorigenesis, initiation by modified Ha-*ras* DNA, 163

U

Undesired effects, systemically active toxicants, 41
Urban application, pesticide, 189–208
Urinary creatinine, excretion rate, 197
Urine
  direct measurement of toxicant or metabolite, 124–126
  occupational studies, 125

U.S. Air Force, Operation Ranch Hand, 126–128

V

Validation, biological markers, 113–115
Variability, marker selection, 112
Viability index, reproductive toxicity tests, 76–77
Vietnam
  exposure to Agent Orange and other defoliants, 126–128
  teratogenesis of 2,4,5-T, 81
Vinyl chloride
  angiosarcoma, 2, 4
  carcinogenesis, 154
Vitamin D, inhibition by cadmium, 178
Vitamin D metabolism, in utero exposure to lead, 6

W

Warranty of confidentiality, death certificate from South Carolina, 240
Whole-embryo systems, 73–76
Worker surveillance, indicator of health risk, 153–173
World Health Organization, chemicals that exhibit toxicity thresholds, 60

X

Xenobiotic metabolism, detoxification process in liver, 45

Z

Zinc refinery, cadmium contamination, 179

Production: Margaret J. Brown
Acquisition: Cheryl Shanks
Indexing: Colleen P. Stamm
Cover design: Jack Ballestero

Typeset by Compset Inc.
Printed by United Book Press, Inc, Baltimore, MD
Bound by American Trade Bindery, Baltimore, MD

Bestsellers from ACS Books

The ACS Style Guide: A Manual for Authors and Editors
Edited by Janet S. Dodd
264 pp; clothbound ISBN 0–8412–0917–0; paperback ISBN 0–8412–0943–X

The Basics of Technical Communicating
By B. Edward Cain
ACS Professional Reference Book; 198 pp;
clothbound ISBN 0–8412–1451–4; paperback ISBN 0–8412–1452–2

Chemical Activities (student and teacher editions)
By Christie L. Borgford and Lee R. Summerlin
330 pp; spiralbound ISBN 0–8412–1417–4; teacher ed. ISBN 0–8412–1416–6

Chemical Demonstrations: A Sourcebook for Teachers,
Volumes 1 and 2, Second Edition
Volume 1 by Lee R. Summerlin and James L. Ealy, Jr.;
Vol. 1, 198 pp; spiralbound ISBN 0–8412–1481–6;
Volume 2 by Lee R. Summerlin, Christie L. Borgford, and Julie B. Ealy
Vol. 2, 234 pp; spiralbound ISBN 0–8412–1535–9

Chemistry and Crime: From Sherlock Holmes to Today's Courtroom
Edited by Samuel M. Gerber
135 pp; clothbound ISBN 0–8412–0784–4; paperback ISBN 0–8412–0785–2

Writing the Laboratory Notebook
By Howard M. Kanare
145 pp; clothbound ISBN 0–8412–0906–5; paperback ISBN 0–8412–0933–2

Developing a Chemical Hygiene Plan
By Jay A. Young, Warren K. Kingsley, and George H. Wahl, Jr.
paperback ISBN 0–8412–1876–5

Introduction to Microwave Sample Preparation: Theory and Practice
Edited by H. M. Kingston and Lois B. Jassie
263 pp; clothbound ISBN 0–8412–1450–6

Principles of Environmental Sampling
Edited by Lawrence H. Keith
ACS Professional Reference Book; 458 pp;
clothbound ISBN 0–8412–1173–6; paperback ISBN 0–8412–1437–9

Biotechnology and Materials Science: Chemistry for the Future
Edited by Mary L. Good (Jacqueline K. Barton, Associate Editor)
135 pp; clothbound ISBN 0–8412–1472–7; paperback ISBN 0–8412–1473–5

For further information and a free catalog of ACS books, contact:
American Chemical Society
Distribution Office, Department 225
1155 16th Street, NW, Washington, DC 20036
Telephone 800–227–5558